APPLIED DESCRIPTIVE GEOMETRY

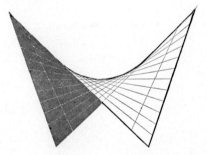

Fifth Edition

APPLIED DESCRIPTIVE GEOMETRY

Frank M. Warner
  Professor Emeritus of General Engineering
  University of Washington

Matthew McNeary
  Professor of Engineering Graphics
  Head, Department of Engineering Graphics
  University of Maine

McGRAW-HILL *Book Company*
New York   Toronto   London   1959

*This book was set in Linotype Primer, a type
face designed by Rudolph Ruzicka in 1954. The
heads are Spartan Heavy and the chapter titles
Venus Bold Extended.*

PREFACE

The purpose of this book is to teach the analysis and solution of three-dimensional problems through application of the principles of orthographic, or multiview, projection. Descriptive geometry differs from basic engineering drawing in that it is not concerned primarily with shape description for the purpose of communication between designer and manufacturing craftsman, but rather is concerned with the analysis of space relationships that precedes design. This analysis is accomplished for the most part by viewing the object or situation in such a way that the relationship being investigated is revealed and made measurable. Practically, this is achieved by drawing successive auxiliary views from any two given views until the desired results are apparent.

A review of orthographic projection and auxiliary-view technique is presented first, and following this, point, line, and plane problems are introduced and solved by the auxiliary-view method. A chapter on revolution shows how some problems can be solved advantageously by rotating the object rather than changing the point of view of the observer. In addition to demonstrating many important applications of descriptive geometry to conventional engineering fields, graphical methods in structural geology are described, and the graphical manipulation of vectors with particular attention to the analysis of concurrent noncoplanar forces in structures is treated in some detail. The usual problems on curved lines and surfaces are introduced with emphasis placed on cones and cylinders because they occur most frequently in engineering practice.

The last two chapters are devoted entirely to exercises and problems. Chapter 8 has 424 practice exercises arranged in 66 groups. As each principle is introduced in the text, reference is made to one of these groups, with corresponding practice exercises. These are given without data, for working freehand on the blackboard or at home, and the intention of the authors is that these be checked for method only.

Chapter 9 is a very important part of the text, because it furnishes the opportunity to apply principles in a practical way. These problems are carefully laid out ready to assign to a class. They have all been tested, and many of them will force the student to do some clear thinking. The solutions should be checked carefully for correctness and clearness of work. By altering the data slightly, the problems in this chapter will be found to be sufficient for several years' use, and they should also furnish a basis for many new ones that may be taken from the experience of any wide-awake instructor.

It has been found that students complete about 30 per cent more problems by using the Warner and Douglass Problem Book which accompanies this text. This book contains many new problems partially laid out to save student time.

The fifth edition of the text has been prepared with the intention of retaining all the features that have made this book so attractive to teachers and students in the past and with every effort to reinforce its strong points: conciseness, readability, and challenging applied problems.

Upon the suggestion of many teachers of descriptive geometry, several changes have been made in the interests of clarity and substantial additional coverage has been provided in subjects where a need was expressed. Definitions and theorems have been slightly modified and more fully illustrated. Sections have been added on visibility, shortest level line and shortest line of given slope between two lines, contour-map problems, geology applications, coplanar vectors, and the fairing of complex double-curved surfaces. The portion of the chapter on revolution dealing with dihedral angle and the angle between a line and plane has been altered to present a solution for a general case.

All illustrations in Chapter 1 to 7 inclusive have been redrawn and many new ones added. Notations on illustrations have been set in Venus Medium Extended type, a style that most nearly approaches engineering freehand lettering.

Several new applied problems have been added in Chapter 9, with most of them inserted in the early problem sections to offer a wide variety of applications of first principles. These have been placed at the end of each section so that the numbering of familiar problems remains unchanged.

The new co-author considers it a privilege to work with Prof. Frank M. Warner, gentleman, master teacher, and pioneer in the movement to make descriptive geometry a live and useful resource for the modern engineer.

Thanks are extended to Prof. E. R. Weidhaas for assistance with the illustrations and to Prof. E. A. Kelso for help with proofreading.

M. McNeary

CONTENTS

1

ORTHOGRAPHIC DRAWING

1.1 Introduction

Orthographic drawing is one of the most important subjects in the entire engineering curriculum. It bears the same relation to the engineering profession as the subject of English bears to our daily life; it is the means by which engineers communicate ideas. When it is properly used, this language of drawing is understood at once by engineers of all nations.

Drawings are not only convenient to the engineer; they are practically indispensable. A single drawing may sometimes contain a sufficient number of ideas to require an entire volume for their adequate expression in words.

Most drawings are made for the purpose of transmitting an idea to someone else. The draftsman who makes the drawing should, first of all, be able to visualize the object he wishes to draw, and his mental picture of this object should be perfectly clear. He must also have the ability to record this mental picture or idea on paper as a drawing, in such form that it will be perfectly clear to anyone skilled in the art of reading drawings.

Other drawings are made for the purpose of making certain graphical determinations in the solution of a space problem. These are more in the nature of layout drawings and are not always made for the purpose of transmittal to other people for reading. They may be made for the sole purpose of solving a calculation problem graphically.

A well-trained engineer is thoroughly familiar with both kinds of drawings. He knows how to make proper commercial drawings, and he knows how to solve any drafting-board problem quickly and correctly. He also appreciates the work involved in making a drawing and the limits of accuracy which may be reasonably expected. *Above all, he knows how to read drawings, because every job with which he will ever come in contact will be designed, contracted for, built, inspected, bought, sold, operated, or maintained through the medium of drawings.*

The principles of drawing and the drafting problems presented in this text are intended to give a logical training and a practical experience in orthographic drawing. A thorough understanding of these pages develops in the student confidence in his ability to visualize space problems, to make proper orthographic drawings, to construct a correct graphical solution of any space problem, and to read and to understand all types of drawings.

1.2   Descriptive geometry

Descriptive geometry is the subject which teaches how to make graphical solutions of space problems by the use of the same principles of orthographic drawing that are used in making simple views of an object. More briefly, descriptive geometry is orthographic drawing applied to the solution of more advanced space problems. The two should not be considered to be two different subjects which require the use of different tools or of different thinking processes. The same elementary principles which are used by an engineer to completely describe an object by orthographic views may be used to solve the more complicated problems in descriptive geometry. The reader should understand, at once, that descriptive geometry, as it is taught in this text, furnishes a method of solution for many practical problems, which requires the use of the identical principles he has already learned in his elementary drawing.

The material to be presented cannot be thoroughly understood without a proper knowledge of elementary drawing. There are many methods for teaching orthographic drawing, all of which have the same fundamental basis but differ in a few minor respects. In order that all readers of this text may become familiar with the conception of orthographic drawing upon which the whole book is based, the rest of this chapter is devoted to an explanation of orthographic views and the relationships which exist between them.

1.3   Change-of-position, or direct, method

If it is desired to draw different views of an object, the draftsman first imagines it to be placed in some definite position. This position is usually the one the object would naturally occupy. An engine base, a bridge, a truss, or a column footing would always be imagined to be resting in its natural position. Small castings, however, may be imagined in any desired position, in order to have the views of the drawing show the object in the easiest way and to the best advantage. After the object has once been placed in some definite position, the draftsman never imagines it to be moved or turned around. If he wishes to see a different side, he simply imagines himself walking around it in order to occupy a different position in space. The observer changes his position in order to look directly at the portion of the object which he wishes to see. Thus, this method is called the change-of-position, or the direct, method of drawing.

1.4   Definitions

1. *Orthographic projection* means "right-angle projection" and is a method of drawing which uses parallel lines of sight at right angles to an image plane.

2. A *line of sight* is a straight line from the eye of the observer to a point on the object. Since all lines of sight for a given view are parallel, the eye of the observer is either at an infinite distance away or it occupies a different position when looking at each point on the object.

3. The *image*, or *picture, plane* is the plane on which an orthographic view is projected. It is always perpendicular to the lines of sight for any view, and it is always between the observer and the object.

4. A *top*, or *plan, view* is an orthographic view for which the lines of sight are vertical and for which the image plane is level.

5. An *elevation view* is an orthographic view for which the lines of sight are horizontal and the image plane vertical.

6. A *folding line* is the line of intersection between two image planes and is the line one image plane is folded on to bring it into the plane of the other image plane. A folding line may be considered an edge view of an image plane when the image plane is folded back 90° from the plane being viewed, and in this sense it

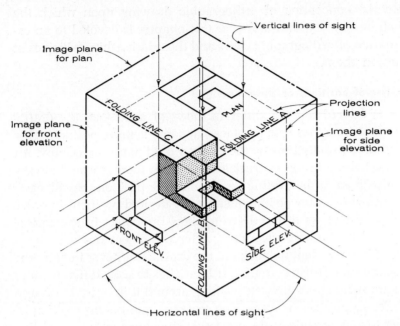

Image plane for plan

Image plane for front elevation

Vertical lines of sight

FOLDING LINE C

PLAN

FOLDING LINE A

Projection lines

Image plane for side elevation

FRONT ELEV.

FOLDING LINE B

SIDE ELEV.

Horizontal lines of sight

FIG. 1.1. Pictorial view illustrating the definitions.

is sometimes called a reference plane. The terms *folding line* and *reference plane* are interchangeable.

7. *Projection lines* are straight lines at right angles to the folding line which connect the projection of a point in one view with the projection of the same point in another view. They are very necessary for obtaining views, but they are not always shown on a drawing.

Figure 1.1 is a pictorial drawing illustrating all of these seven definitions so as to give a correct conception of their relations to each other in space. The object itself is shown shaded and behind all three image planes. Attention is called to the fact that each view of the object is right on the image plane for that view. In other words, the image plane is the plane of the paper itself on which the actual drawing is made. A most careful study of the foregoing definitions and illustration should be made.

1.5   **Folding the image planes**

All drawings in practice, including all views, are actually made in one plane, which is the plane of the paper or drawing board. Hence the image planes, which are shown in Fig. 1.1, must all be brought into the same plane in order to occupy the position where the views are actually drawn by the draftsman. This may be done in two different ways.

The first way to fold the image planes is to imagine the plan

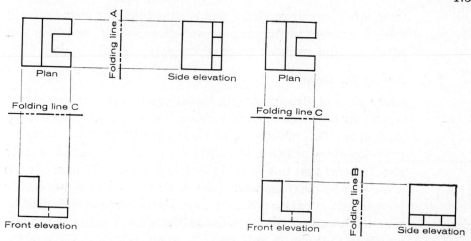

FIG. 1.2. The views folded into the plane of the plan.   FIG. 1.3. The views folded into the plane of the front elevation.

image plane to remain stationary and the other two image planes to be folded about the folding lines A and C until they lie in the same plane as the plan image plane. This will bring the three views into the positions shown in Fig. 1.2, with the side elevation projecting from the plan view.

The second way to fold the image planes is to imagine the front image plane to remain stationary and the other two to be folded about the folding lines B and C until they lie in the same plane as the front image plane. This will bring the three views into the position shown in Fig. 1.3, with the side elevation projecting from the front elevation.

In either of these methods the resulting views themselves are identical, but the side elevation simply occupies a different position on the paper. Both methods are absolutely correct and both are largely used by draftsmen on various kinds of commercial drawings. The author favors the first method for beginners, for the simple reason that the plan view seems to be the *basic view.* The plan view has vertical lines of sight; these can have only one possible direction, and therefore there can be only one possible plan view. It is assumed that vertical lines of sight always look down and not up, in accordance with standard commercial practice. Lines of sight for elevation views are level and may have an infinite number of directions. Therefore we may have an infinite number of elevation views of any object. Since, then, there can be only one plan view, but any number of elevation views, it seems logical to treat the plan as the basic view and to take all elevation views from the plan. Experience has shown, too, that it is easier for most students, at first, to take elevation views from the plan view. After more experience has been gained, the view that will solve a problem in the simplest and clearest way, or which will better suit the available space on the paper, should be used.

1.6   Placing the views

Every piece of structural work, small casting, or machinery part which an engineer intends to build must first be drawn by a draftsman. This drawing must be an accurate and complete description of the article to be built, both as to its shape and as to its size. In order that this drawing may be interpreted in exactly the same way by everyone who reads it, there must be a universally recognized system for placing views. The system which is used by practically every drafting room in the United States is the one which is followed in this text.

In this system the image plane is always considered to be between the observer and the object, as has been shown in Fig. 1.1. If this rule is adhered to and if the image planes are folded as explained in Section 1.5, the drawing automatically becomes a third-angle drawing, which conforms to the best commercial drafting-room practice in our country. Without explanation of the four-quadrant conception of drawing, which is fast becoming obsolete, a third-angle drawing may be defined as one in which a view is always placed on the same side of an object as the observer of the object. The explanation which has just been given is illustrated in Fig. 1.4.

An observer, who imagines himself standing south of the house to view it, places this view on the south side of the plan, that is, on the same side from which the object was viewed. This is the most natural placement for any view, and *this method should be strictly adhered to* in all drawing problems.

As an aid in understanding this arrangement more clearly, it is most helpful to copy the drawing of Fig. 1.4 on a sheet of paper and to cut out the two corners on the dashed lines as indicated. Then fold the image planes for the three elevation views down into their proper positions in space, until they are at right angles to the plan image plane. Then stand these image planes on a table and walk around them, keeping your lines of sight level. Imagine the house itself to be in behind the four image planes. Notice whether or not the elevation views seem to be correctly placed.

1.7   Distance from the folding line

In Fig. 1.4, special attention is called to three self-evident facts which are most important:

1. The highest point on the house, or the ridge, is always toward the plan in all elevation views.

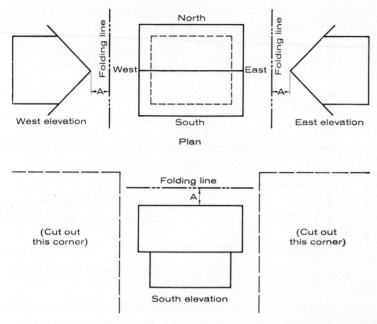

FIG. 1.4. Accepted method of placing views.

2. The house is the same height in all elevation views.

3. In all elevation views the ridge is the same distance below the folding line. This was to be expected, since the ridge itself, in space, is the distance A below the plan image plane and hence will appear the distance A below the folding line (which is the plan image plane appearing as an edge) in any view having level lines of sight. This being true for these three elevation views, it is likewise true for all elevation views. It is also true for any other point on the house, as well as for the ridge. These facts may be stated more clearly in the form of a rule.

Rule 1. Any point on an object will appear the same distance below the folding line in all elevation views that are related to the plan.

1.8   Notation

The solution of more difficult problems in drawing often requires the use of more than two or three views. A very simple system of numbering the views and folding lines is a great aid to the draftsman in keeping the drawing clear, and it eliminates the necessity of labeling the views. Since the plan is considered to be the basic view it is always given number 1. The next view to be drawn is given number 2, and all other views are then numbered in the order in which they are drawn, as shown in Fig. 1.5. The numbers of the views are placed on each side of the folding line as shown. The folding line, therefore, carries the numbers of the views it lies between. In some problems it is also convenient to

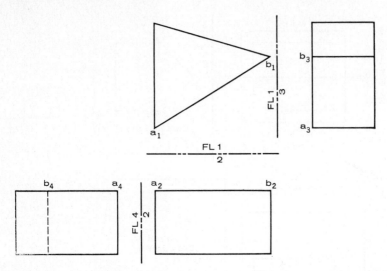

FIG. 1.5. Notation system.

letter each point in each view. Lower-case letters are always used and they are given a subscript number which is the same as the view number. Anyone who is familiar with this notation can tell at a glance which is the plan view, and what the order of solving the problem was, by the sequence of the view numbers.

Note: It is sometimes desirable in stating a problem to give the front and side elevations without the plan view. In this case, when only views 2 and 3 are given, it always should be assumed that view 2 is the *front* elevation and the plan view is directly above it.

Capital letters are used in referring to a line itself in space, as the corner *AB* of the wedge in Fig. 1.5.

The folding lines are shown with dashes, one long and two short. They are shown as rather heavy lines in this text, in order to distinguish the views better. However, for accurate pencil solutions, the folding line should be drawn as a fine line but with the same notation.

The notation to be used for the object lines and solution lines is not specified, because an exact specification becomes too complicated and reduces the opportunity for the student to exercise his own judgment. In general, it may be said that construction lines should always be very fine and accurate; the object lines should be more pronounced. Judgment must be used to make the drawing clear and understandable at a glance.

1.9   Practice problems

See Chapter 8, Group 1.

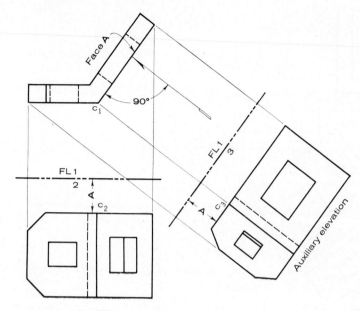

FIG. 1.6. Auxiliary elevation.

1.10   Auxiliary elevation views

The statement has already been made that elevation views have level lines of sight and that level lines can have an infinite number of directions, making possible an infinite number of elevation views. The common elevation views are the front, the rear, the right-side, and the left-side views. Any other view using level lines of sight is called an auxiliary elevation view. Often it is necessary to draw such a view in order to see a definite part of an object in its true size and shape. By use of the principle of Rule 1 and the fact that the lines of sight for any view are parallel, any possible auxiliary elevation view is as easily drawn as a side elevation view.

In Fig. 1.6, it is desired to obtain a view of the bracket with lines of sight perpendicular to the face A as shown by the arrow. Then all the lines of sight for this view are parallel to this arrow. The point C on the object, in this view, will appear on a definitely fixed line of sight, or projecting line, and it must be A distance below the folding line, giving c_3. All other points on the object are obtained in this view in the same manner.

Any other auxiliary elevation can be drawn as easily by locating each point on its proper projecting line at the correct distance below the folding line. If any difficulty is experienced in visualizing any part of this explanation, a paper drawing should be made and cut out and folded as suggested for Fig. 1.4.

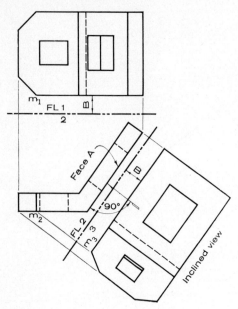

FIG. 1.7. Inclined view.

1.11 Practice problems

See Chapter 8, Group 2. (See Appendix A.1 for method of drawing parallel and perpendicular inclined lines.)

1.12 Inclined views

An inclined view is any view for which the lines of sight are neither vertical nor horizontal. In other words, the lines of sight for
any inclined view must be sloping or inclined.

The object in Fig. 1.6 might have been placed in a different position for purposes of drawing, as in Fig. 1.7. Attention is called to
the fact that in Fig. 1.7 the view marked 2 is a front elevation
view and *not* the plan, although it is identical with the plan of
Fig. 1.6. The position an object occupies in space must always be
kept clearly in mind. The face A is now an inclined plane and, in
order to look with lines of sight perpendicular to it, inclined lines
of sight will have to be used which are parallel to the arrow. The
point M on the object lies in a fixed line of sight. But, in this
case, the distance B from the folding line is obtained from the
plan view because the point M is actually B distance behind the
front image plane. In both the inclined view and the plan the folding line is the front image plane appearing as an edge. This determines m_3 in the inclined view. All other points on the object
are located in the inclined view in a similar manner. By this
method an inclined view is drawn as easily as any elevation view.

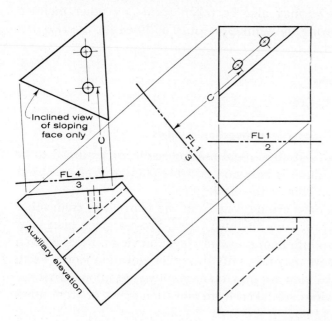

FIG. 1.8. Inclined view from auxiliary elevation.

Once more the drawing should be made and the paper cut and folded, that there may be a clear understanding that the *B* distance is the correct one to use.

1.13 Practice problems

See Chapter 8, Group 3.

1.14 Additional inclined views

In Fig. 1.7 the inclined view was taken directly from the front elevation. Sometimes it becomes necessary to draw an inclined view from an auxiliary elevation, as in Fig. 1.8. The method is just the same as that shown in Fig. 1.7, but in this case the distance *C* from the folding line in the inclined view is taken from the plan view as shown on the drawing. It is an entirely different measurement from the *B* distance of Fig. 1.7. In this case it is taken from the plan as shown because the center of the nearest hole on the object is actually *C* distance behind the image plane for the auxiliary elevation.

The drawing should again be made, and the paper cut and folded, in order to see clearly why this *C* distance is the correct one to use. However, if the drawing is even turned, temporarily, so that the auxiliary elevation occupies the position of the front elevation, it will be very evident why the *C* measurement is correct.

Inclined views may also be taken from any other inclined views by following the same procedure outlined for the two preceding problems.

1.15 Practice problems

See Chapter 8, Group 4.

1.16 Summary of all possible orthographic views

1. **Plan view.**   If all vertical lines of sight are assumed to be looking down, there is only one possible plan view of an object fixed in space. This is the *basic view.*

2. **Elevation views** are derived from the plan view, from other elevation views, or from inclined views. From a given plan view we may take an infinite number of elevation views. From a given elevation view we may take only two other elevation views. From a given inclined view we may take only two elevation views.

3. **Inclined views** are taken from elevation views or from other inclined views, but *never* from a plan view. From a given elevation view we may take an infinite number of inclined views. From a given inclined view we may take an infinite number of inclined views.

Figure 1.9 illustrates all the different possible orthographic views which have just been explained. The arrows indicate the

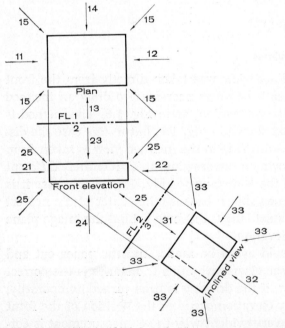

FIG. 1.9. Lines of sight for all possible orthographic views.

different possible directions for lines of sight, and each arrow is numbered and referred to in Table 1. This table shows clearly what kind of view would be obtained by using lines of sight parallel to each numbered arrow shown in Fig. 1.9. After carefully studying Table 1 and Fig. 1.9, one should always be able to name correctly any view which is taken from any other view.

Table 1

View taken from		Arrow numbers	View obtained	
Plan	(1)	11	Left-side	elevation
		12	Right-side	elevation
		13	Front	elevation
		14	Rear	elevation
		15	Auxiliary	elevations
Front elevation	(2)	21 (Same as 11)	Left-side	elevation
		22 (Same as 12)	Right-side	elevation
		23	Plan	
		24 (Seldom used)	Bottom	view
		25	Inclined	views
Inclined view	(3)	31 (Same as 13)	Front	elevation
		32 (Same as 14)	Rear	elevation
		33	Inclined	views

1.17   Related views

Views are said to be *related* to each other when:

1. Their image planes make 90° with each other in space.
2. They have folding lines between them.
3. The two views of any point lie on the same projecting line which is at right angles to the folding line between the views.

All three of the foregoing conditions must be satisfied if two views are related. Views which are related to a common view are not related to each other, but they do have a common dimension which is at right angles to the folding line in both views. In Fig. 1.10, views 1 and 4 are related to view 2. But view 4 is not related to view 1, because corresponding points on the object will not project in parallel lines between these views and because there is no folding line between these two views.

The drawing of Fig. 1.10 gives a complete summary of the relationships between the various possible views. Table 2 shows just which views are related to each other. For instance, the only views that are related to the plan view 1 are views 2, 3, and 5, all of which must be elevation views.

The correct measurements to take in order to obtain any new view are also shown in Fig. 1.10. They may be summarized in another rule.

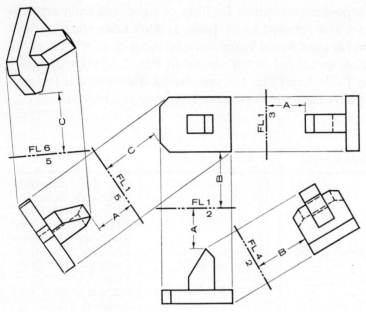

FIG. 1.10. Related views.

Rule 2. In all views that are related to a common view, the object, or any point on the object, is the same distance away from the folding line.

If the finger is placed on any orthographic view, all views which project from it (or are related to it) are the same distance from the folding line. This fact may easily be checked by placing a

Table 2

Views no.	Are related to	
	No.	Name
2-3-5	1	Plan
1-4	2	Front elev.
1	3	Side elev.
2	4	Inclined
1-6	5	Auxiliary elev.
5	6	Inclined

finger on some view, such ás view 5. Views 1 and 6 are both related to view 5 and they both lie C distance away from the folding line.

Another method of checking the measurement from the folding line is temporarily to consider any view as a plan. Then all views related to that view are, temporarily, elevation views, and the measurement from the folding line seems more easily checked. However, this last method is not recommended except in cases of extreme difficulty in visualization.

1.18   Purpose of folding lines

The use of image planes and folding lines which has just been explained is the most fundamental method for clearly visualizing the relationship between views and is the proper way to measure the size of an object in a new view which is to be drawn. In shop drawings and detail drawings of simple objects which require only two or three views for describing them completely, the folding line is not usually shown, although it may be imagined to be there. For such drawings it is perfectly proper and it is good commercial practice to omit the folding line entirely.

However, in the more complicated problems of descriptive geometry which follow in this text, it is imperative that there be some place from which to make measurements and some standard way of measuring. The folding-line system provides this means of measuring and is therefore followed throughout this text. It should be used by the student in all the problem solutions, for it will prove to be very convenient and will be a very great help in making the solutions.

1.19   Visualization

It is highly important to establish the *habit of visualizing* the position of a line or plane in space by looking at the related views of that line or plane. Two views of a line *are only the way some definite line in space appears* when viewed from two different positions. The views are only the *means* of showing on paper the exact position of some actual line in space. *Attention should always be focused on the line or object itself* as pictured by the given views.

Figure 1.11 shows two views of four lines having different positions in space. Cover up the front elevation and look at the plan view only. Lines 1 and 2 appear to have the same direction; lines 3 and 4 appear to have the same direction. All four lines might be level, or they might slope from *A* toward *B* or from *B* toward *A*. From the plan view it is *impossible* to determine their positions in space because the third dimension (or elevation) cannot be shown in this view.

Now look at the front view in connection with the plan. For line 1, *A* and *B* are at the same elevation, but for line 2, point *A* is higher than *B*. Therefore line 1 is level, but line 2 slopes down

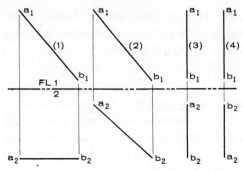

FIG. 1.11. Visualization of lines.

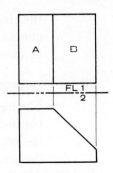

FIG. 1.12. Visualization of planes

from *A* toward *B*. In the same way notice that line 3 slopes down from *A* toward *B* but line 4 slopes up from *A* toward *B*. The student should hold his pencil in front of him in exactly the way that each of the four lines lies in space. In any future problem two views of a line should indicate to him immediately just how that line lies in space.

Figure 1.12 shows two views of an object which is made up of lines and plane surfaces. Again it is impossible to tell from the plan view alone whether surfaces *A* and *B* are level or sloping. But the front view shows clearly that surface *A* is level and surface *B* is sloping. Thus surfaces must be read and located in space by reference to more than one view.

This is the basis for all visualization, and the secret of it is *always to read two (or more) related views together to get complete space information.*

1.20   Reading orthographic views

The ability to read drawings is a necessary and valuable attribute which should be cultivated by a young student. He should learn to look at three related views of an object and to form an exact men-

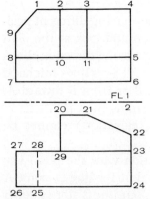

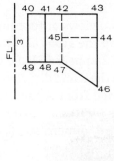

FIG. 1.13. Reading orthographic views.

tal picture of that object. To learn to do this, he should practice by selecting any line or plane surface on an object in one view and locating it in another view, and hence definitely in space as in Section 1.19. The best way to get this practice is to number points on each view which define lines or surfaces in that view. This is illustrated in Fig. 1.13 where an object is shown in three views. The numbers, like 3-4-5-11, do not necessarily locate points on the object, but they indicate that in that view some surface on the object appears to be 3-4-5-11, although elevations are unknown. Glancing at the front elevation, by projection the only surface it could be on the object is 21-22. Therefore the highest point on that surface is 21, which would be line 40-49 in view 3, and the lowest point is 22, which would be line 41-48 in view 3 by folding-line measurement. Therefore the surface 3-4-5-11 in the plan must be 40-41-48-49 in view 3 by both projection and measurement.

The following hints should be studied and followed:

1. A line at right angles to an image plane for any view will appear as a point in that view. Line 45-44 is a point in view 1.

2. Any plane surface will either have a similar shape or will appear as a line in related views. Surface 3-4-5-11 appears in a similar shape (40-41-48-49) in view 3 and as a line (21-22) in view 2.

3. Any line in any view indicates a change of direction between two adjacent surfaces. In the plan view the line 3-11 *guarantees*

Table 3. Locating lines

Find this line	In this view	Record answers here
9	3	45-44
9	2	
41-43	2	22-24
41-43	1	
5-8	3	
5-8	2	
6-7	2	
6-7	3	
25-28	1	
25-28	3	
22	1	
22	3	

Table 4. Locating planes

Find this surface	In this view	Record answers here
2-3-11-10	2	20-21
2-3-11-10	3	
7-8-9	3	45-44-46-47
7-8-9	2	
3-4-5-11	3	
3-4-5-11	2	
40-42-47-49	1	
40-42-47-49	2	
41-43-46-47-48	2	
41-43-46-47-48	1	
42-43-44-45	1	
42-43-44-45	2	
46-47	1	
46-47	2	
42-47	1	
42-47	2	

that surfaces 2-3-11-10 and 3-4-5-11 are different surfaces and not one continuous plane.

4. To find a surface in other views, remember that it must project correctly and that distances from the folding line must also check in views related to the same view.

The student should now fill in (in pencil) the correct answers in Tables 3 and 4 on page 17. After a little practice he will read drawings very rapidly without using any numbering system.

2

FUNDAMENTAL AUXILIARY VIEWS

2.1 Necessity for auxiliary views

The preceding chapter has shown that it is possible for an observer to stand at any desired place in order to view an object at any desired angle. It has also shown how the draftsman draws the different views on paper and keeps them properly related. The reason why it sometimes becomes necessary to view objects or problems from different angles will now be explained.

All commercial drawings must be dimensioned. Almost every object which is drawn is bounded by lines and planes. No line or plane can be dimensioned in any view unless that line or plane appears in its true size in that view. No work line on a steel drawing can be dimensioned in any view except in the view which shows its true length. No angular cut on a steel plate can be dimensioned unless shown in its true size. Therefore the draftsman must know how to obtain quickly that view of an object which will show it exactly the way he must see it in order to dimension it properly.

2.2   Four fundamental views

The experience of the author and of many other engineers has proved to their entire satisfaction that the necessary views, as explained in Section 2.1, may always be obtained by drawing certain basic or fundamental views. Practically any object composed of lines and planes that an engineering draftsman would ever have to draw can be drawn and dimensioned by the use of one or more of the four fundamental views listed below.

View A.   Showing the true length of a straight line.
View B.   Showing a straight line as a point.
View C.   Showing a plane as a line or an edge.
View D.   Showing a plane in its true size and shape.

In order fully to explain these four views, the fundamental conceptions of lines and planes must be considered first.

2.3   Lines: definitions

1. A *line* is the path of a moving point.

2. A *straight line* is the path of a point moving constantly in the same direction. Hereafter in this text it may be assumed that the word "line" implies a straight line unless otherwise designated. Since any two points determine a straight line, a line is usually designated by its two extremities if it has a fixed length. Otherwise any two points on the line may be chosen at random for the purpose of locating the entire line in any other view.

3. A *level line*, or a *horizontal line*, is a line on which every point lies at the same elevation. See Fig. 2.1a.

4. A *frontal line* is a line which lies parallel to the image plane for the front elevation. It may be level, vertical, or inclined and it always shows in its true length in the front elevation. See Fig. 2.1b.

5. A *profile line* is a line whose plan and front elevation views are both perpendicular to the folding line. Its true length and slope can be seen only in a side elevation view which is sometimes called a profile view. See Fig. 2.1c.

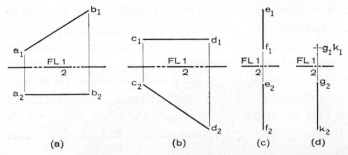

FIG. 2.1. Types of lines: (a) level line, (b) frontal line, (c) profile line, and (d) vertical line.

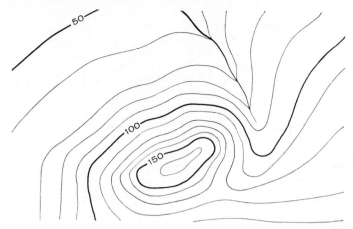

FIG. 2.2. Contour lines on a map.

6. A *vertical line* is one that is perpendicular to a level plane. A line that is perpendicular to an inclined plane is not vertical. *Vertical* and *perpendicular* have quite different meanings. See Fig. 2.1*d*.

7. A *contour line* on a map is a line, straight or curved, connecting a series of points which are at the same elevation or level. It is therefore a level line. Figure 2.2 shows a map with 10-ft contour lines, which means that the contour lines are shown at every difference of 10 ft in elevation. A person who is at the 80-ft level and who wishes to walk around remaining at the same elevation would have to follow the 80-ft contour line.

2.4   Bearing

The bearing of a line is the angle between its plan view and the plan view of a line running north and south. The angle less than 90° is usually given for the bearing, and it may be given from either the north point or the south. If the bearing of a line, from a given starting point, is said to be N60°E, it means that the plan view of the line is turned away from the north line 60° toward the east, as is shown in Fig. 2.3. The bearing of a line is not at all affected by its slope. If the bearing of a line is given as N75°W, the direction of the plan view is absolutely fixed regardless of whether the line is level or whether it has a steep slope. The bearing of any level line on the plane of a vein of ore is called the "strike." See Section 3.44.

2.5   True length of a line: view A

A line must lie at right angles to the lines of sight for any view in order to appear in its true length in that view. In other words, a line must lie parallel to the image plane for the view which shows its true length. Only level lines appear in their true length in the

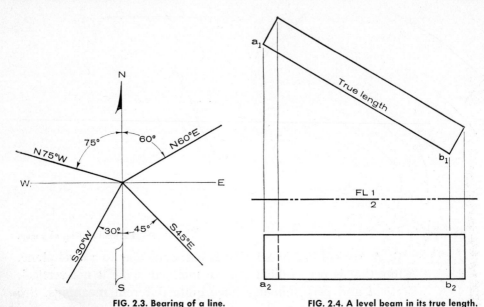

FIG. 2.3. Bearing of a line.   FIG. 2.4. A level beam in its true length.

plan view, because the plan image plane is level and all lines that are parallel to it must also be level. In Fig. 2.4 the front elevation shows that the timber is in a level position and parallel to the plan image plane. Therefore the plan view shows the true length of this timber.

Any vertical line appears in its true length in all elevation views, because the image planes for all elevation views are vertical and hence any vertical line is parallel to them all. In Fig. 2.5

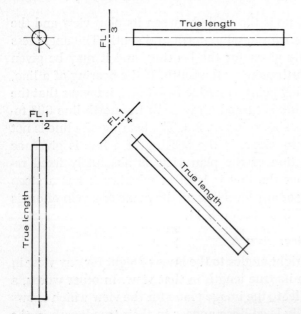

FIG. 2.5. A vertical mast in its true length in all elevation views.

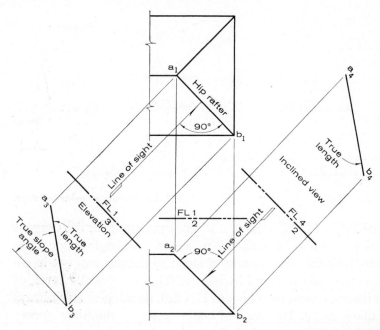

FIG. 2.6. True length and true slope of a hip rafter.

the vertical mast shows in its true length in all three elevation views because it is parallel to the image planes for all three.

In Fig. 2.6 are shown a plan and a front elevation of a portion of a hip roof. The hip rafter *AB* does not show in its true length in the plan because, by inspection, it is not level. It also does not show in its true length in the front elevation, because the lines of sight for that view are not at right angles to it. If its true length is desired, a new elevation view must be drawn with lines of sight at right angles to the rafter. This auxiliary elevation is easily determined by the use of parallel lines of sight and of the principle of Rule 1, Section 1.7. Also an inclined view may be drawn from the front elevation, with lines of sight at right angles to the rafter as shown by the arrow. This inclined view also shows the rafter in its true length, which should check exactly with the true length determined in the auxiliary elevation.

2.6   Practice problems

See Chapter 8, Group 5.

2.7   True slope of a line

The true slope of a line is the angle the line makes with a horizontal plane, as is shown in Fig. 2.7. The per cent grade of a road, considered as a centerline, is 100 times the tangent of the angle between the road and a horizontal plane, as is shown in Fig. 2.8. A road rising a vertical distance of 10 ft in going a horizontal dis-

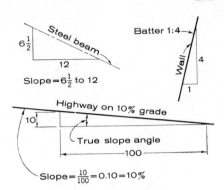

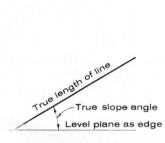

FIG. 2.7. True slope angle.

FIG. 2.8. Methods of indicating slope.

tance of 100 ft is said to have a 10 per cent grade, or the tangent of the slope angle is 10/100. The most common engineering practice is to designate the amount of slope by the per cent grade, although it may be measured in degrees. The slope, or the bevel, of a steel beam is given as shown in Fig. 2.8, the longest side always being taken as 12. The slope of concrete walls which are almost vertical is sometimes called "batter" and may be indicated as shown or by just a lettered note in the plan view. All these methods for dealing with slope will be used in the problems of Chapter 9.

Since the slope angle between any line and a level plane lies in a vertical plane which contains the line, its true size can be seen only when viewed at right angles to that vertical plane, with level lines of sight. Elevation views are the only views for which level lines of sight are used. Therefore the following rule may be stated.

Rule 3. The true angle of slope of any line can be seen only in the elevation view which shows the true length of the line.

In Fig. 2.6, again, the true slope of the hip rafter is seen in the auxiliary elevation view which shows its true length, and it is so marked in this figure.

Caution: An inclined view may show the true length but NEVER *the true slope of a line,* because it cannot show a level plane as an edge.

2.8 Practice problems

See Chapter 8, Group 6.

2.9 A line as a point: view B

If the lines of sight for any view are parallel to a line in space, that line appears as a point in that view. Figure 2.9 shows the line *MN* occupying three different positions in space. In Fig. 2.9*a*

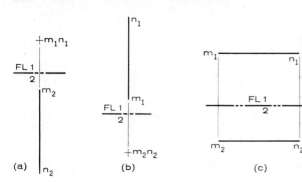

FIG. 2.9. Views showing a line as a point.

the line shows as a point in the plan view because it is in a vertical position. In Fig. 2.9*b* it shows as a point in the front elevation because it is a level line and lies parallel to the lines of sight for the front view. In Fig. 2.9*c* it appears as a point in the side elevation because it is level and lies parallel to the lines of sight for the side elevation.

Figure 2.10 shows four views of a level line *AB*. The new elevation view 4 is drawn from the front elevation, and the line *AB* does not appear as a point in this view. But a new auxiliary elevation, view 3, taken from the plan view does show the line as a point, because the line *AB* is in its true length in the plan. For view 3 the lines of sight are actually parallel to the line *AB* itself in space. For view 4 this is not the case.

Now let the same line *AB* have the end *A* raised a little as in Fig. 2.11. Once more, view 4 is drawn from the front elevation and view 3 is drawn from the plan. But the line *AB* does not show as a point in either of these new views, if the principles of projection have been used correctly. The lines of sight for these new views only *appear* to be parallel to the line *AB* in space. Actually

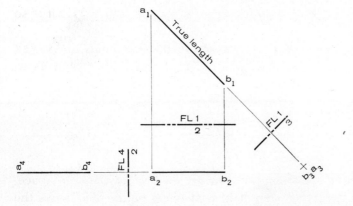

FIG. 2.10. Showing when a line appears as a point.

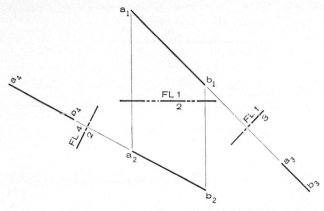

FIG. 2.11. Showing when a line does not appear as a point.

they are not parallel to the line AB, because that line is not in its true length in either view. With the line AB in the position shown in Fig. 2.11 it is impossible to draw any view from the plan or the front elevation which would show it as a point.

Before the line AB can be seen as a point in any view, a new view must be drawn showing it in its true length, as in Fig. 2.12, view 3. An inclined view 4 taken from the true length view 3 is found to show the line as a point. This may easily be proved by actually making the drawing and locating each point (A and B) in the new views by measurement as was explained in Chapter 1. It is self-evident that, since the line AB is inclined, we should have to use inclined lines of sight to see it as a point. This explanation is now summarized in a very important rule.

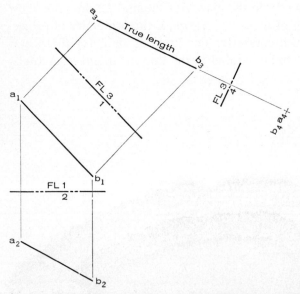

FIG. 2.12. The condition necessary to see a line as a point.

Rule 4. A line must appear in its true length in some view before a view may be drawn showing it as a point.

This rule should be checked by referring again to Figs. 2.9, 2.10, and 2.12, and observing that in every case the view that shows the line as a point is related to the view that represents the line in its true length.

2.10   Practice problems

See Chapter 8, Group 7.

2.11   Planes

A plane is a surface such that, if any two of its points are connected by a straight line, that straight line always lies wholly on the surface; or every point in that line is on the surface.

A plane may be determined in space in three different ways, all of which are common in engineering work and are used in the problems in this text. These three ways are as follows:

1. By any three points not in a straight line, as in Fig. 2.13a.
2. By any two intersecting lines, as in Fig. 2.13b. It should be noted that two intersecting lines have one point in common that projects between views.
3. By any two parallel lines, as in Fig. 2.13c.

A plane which forms a limiting surface of a definite object, like the side of a box, is limited in extent or size. Abstract planes are

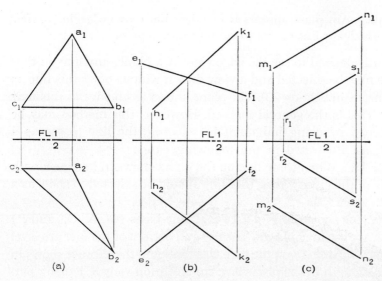

FIG. 2.13. Three ways of representing planes.

usually considered to be indefinite in extent. However, for the purpose of solving problems, *all planes and all straight lines may be considered to be indefinite in extent, even though they may really have a definite size.* A plane or a line, even though it is on a definite object, may be extended beyond the limits of the object if this extension makes the solution easier. The position of the plane or the line in space is not altered by such extension.

A *profile plane* is a plane which shows as an edge in both plan and front elevation views. It is parallel to the image planes for the side elevation views.

2.12   A plane as a line or an edge: view C

In order to furnish a very convincing proof for a very important statement which is next to be presented, it is suggested that the reader perform the following simple exercise.

Place your lead pencil so that it lies flat against your triangle, and hold them both up in any position in front of you keeping their relative position fixed. Now occupy a position such that you can see the pencil appearing as a point and notice that in this view the entire triangle shows as a single straight line. Then try the pencil in some other position on the triangle and see whether the same thing is true again. The triangle represents a plane, and the pencil represents any line lying in that plane. Regardless of where this line lies in the plane, when you see the line as a point the plane appears as a single line or an edge.

This simple illustration of a visual nature gives the next important rule.

> **Rule 5. Any plane appears as a straight line, or an edge, in the view in which any line in that plane appears as a point.**

If it is desired to see the edge view of a plane, any line on that plane may be selected and a view drawn which shows this line as a point, as in Section 2.9. The plane shows as an edge in this new view. This is the general method. However, this method may be shortened by using judgment in selecting the line to see as a point, as illustrated in the following cases.

Case 1.   When some line on the plane shows in its true length in one of the given views, that line should be selected to show as a point.

The pier shown in Fig. 2.14 has four faces (M, N, O, and P) which are inclined planes. The top and bottom of the pier are level planes. The line AB on the face O is level and therefore it shows in its true length in the plan view. An elevation view 3, looking parallel to a_1b_1, shows the line AB as a point. If other points on this

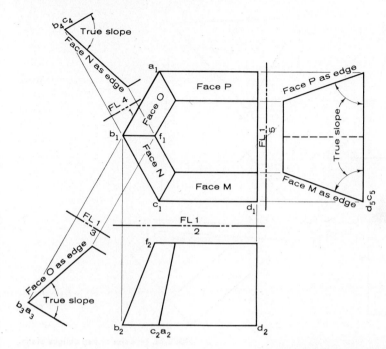

FIG. 2.14. Edge view and true slope of a plane.

face are located in this view, the face will be found to show as an edge, according to Rule 5. In the same way an elevation view 4, looking parallel to the level line BC on the face N, will show that face as an edge. Also an elevation view 5, looking parallel to the line CD on the face M, will show that face as an edge. Also, since the line BF shows in its true length in the front elevation, an inclined view taken from the front elevation and looking parallel to the line BF will show both the face N and the face O as edges.

Case 2.   When no line on the plane shows in its true length in either of the given views, draw a new line on the plane that will satisfy this condition. A level line is the best one to select because its true length shows immediately in the plan view. The prism shown in Fig. 2.15 has a sloping top plane RSM. No line on this plane shows in its true length in either of the given views. A new line MN, which is level, is drawn on the plane in the front elevation and projected to the plan, where it shows in its true length. The new elevation, view 3, is then drawn to show the line MN as a point. The points M, R, and S are all located in this view by Rule 1 and, after they are located, they are found to lie in a straight line. This gives the edge view of the plane RSM as it should according to Rule 5. The frontal line RV could have been selected instead of the level line MN. However, it is advisable to use the level line in most cases.

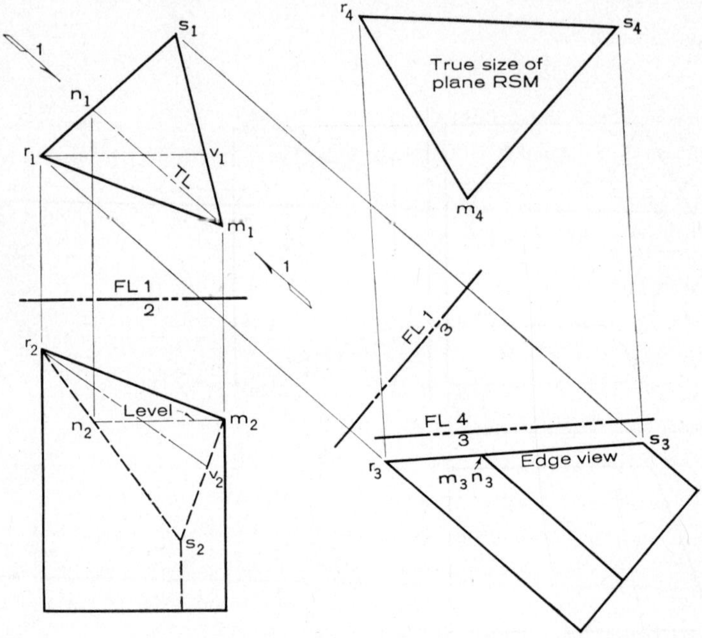

FIG. 2.15. True size of any oblique plane.

2.13 Additional facts about edge views

A very convincing check of the edge view in the preceding problem will be furnished if the student will actually draw several elevation views of the prism of Fig. 2.15. He will soon discover that a thousand auxiliary elevation views could be drawn, and that in only two of these views would the plane *RSM* appear as an edge. These two views would be those for which the lines of sight were in the direction of the arrows numbered 1.

In like manner, just as many views could be drawn from the front elevation. The only ones that will show the sloping plane *RSM* as an edge are the two inclined views whose lines of sight are parallel to the frontal line *RV* on the plane, because a frontal line always shows in its true length in the front elevation.

Attention is called to the fact that, if the plane *RSM* shows in its true size in the inclined view 4, it must lie parallel to the image plane for this view. Therefore all views taken from this inclined view 4 will show the plane *RSM* as an edge and parallel to the folding line, just as it is in the auxiliary elevation view 3. This fact will be useful in analyzing later problems.

2.14 Practice problems

See Chapter 8, Groups 8 and 9.

2.15   True slope of a plane

The angle of slope of a plane is the angle the plane makes with a horizontal plane. It may be measured in degrees or by the tangent of the angle, exactly as the slope of a line may be measured.

> Rule 6. The true angle of slope of a plane can be seen only in an elevation view which shows the plane as an edge.

The horizontal plane must also appear as an edge. This is the reason why the elevation view is specified in Rule 6, *for it is the only view that can show a level plane as an edge.* In Fig. 2.14, again, the true angle of slope of each plane is seen in the auxiliary elevation which shows that plane as an edge. It should be noticed that in each case the angle of slope is the angle between the inclined plane and a level plane.

Caution: An inclined view may show a plane as an edge, but it can NEVER *show its true slope,* since it cannot show a level plane as an edge.

Among mining engineers the true slope of a vein of ore is called the "dip." See Section 3.44.

2.16   Practice problems

See Chapter 8, Group 10.

2.17   A plane in its true size: view D

A plane lying parallel to an image plane, or at right angles to the lines of sight, for any view, shows in its true size and shape in that view. It should be obvious that, in order to see a plane in its true size, one must look with lines of sight at right angles to the plane.

A level plane shows in its true size in the plan view. A vertical plane shows in its true size in the elevation view whose lines of sight are perpendicular to it. But in order to look at right angles to an inclined plane the lines of sight must also be inclined, giving an inclined view.

From the plan and front elevation views of the prism in Fig. 2.15, it would be very difficult to tell, from those two views alone, where to go to look at right angles to the face *RSM.* But in the auxiliary elevation showing this face as an edge it is apparent that the lines of sight must be at right angles to the plane where it appears as an edge, in order to be at right angles to the plane. This idea is also stated in the form of a rule, because it will be used over and over again.

Rule 7. A plane must always appear as an edge before a view can be drawn showing it in its true size.

Figure 2.15 shows how the true size of the face *RSM* is found. The auxiliary elevation view 3 is first drawn, as explained in Section 2.12, in order to show the face *RSM* as an edge. The true size of the face *RSM* is then seen in the inclined view 4 by looking at right angles to this face where it appears as an edge. This inclined view shows all the lines on this face in their true lengths and all the angles on this face in their true sizes. This face may now be completely dimensioned in view 4, and any desired measurement may now be made from this view.

2.18 Practice problems

See Chapter 8, Group 11.

2.19 Summary

Attention is called to the close relationship that exists between these four fundamental views which have just been explained. View *B* cannot be obtained without having view *A* first. View *C* cannot be drawn without having views *A* and *B* first. View *D* cannot be drawn without having views *A*, *B*, and *C* first. In other words, if it is desired to draw a view showing the true size of a plane, a view must first be drawn which shows the plane as an edge. In order to obtain this edge view of the plane, some line on

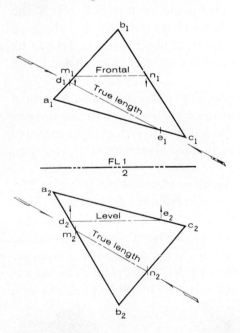

FIG. 2.16. Level and frontal lines.

the plane will have to show as a point in some view. Before a view can be drawn showing a line as a point, there must be a view showing that line in its true length. This last-named view need not always be an extra view, if judgment is used in selecting a line on the plane which already shows in its true length in one of the views already given. The solution is usually simplified by the use of a level or a frontal line on a plane, for these lines show in their true lengths in the given plan and front elevation views, respectively. In Fig. 2.16 any line on the plane parallel to the line MN is a frontal line and will show in its true length in the front elevation. Also any line on the plane parallel to the line DE is a level line and will show in its true length in the plan. The four views having lines of sight parallel to the four indicated arrows will all show the plane as an edge but only the two elevation views will show the slope of the plane.

3

POINT, LINE, AND PLANE PROBLEMS

3.1 Introduction

The common conception of descriptive geometry is that it deals
only with abstract lines and planes. This may be correct from the
standpoint of a mathematician. But descriptive geometry is an
engineering subject and a valuable engineering tool. It was in-
vented by an engineer, Gaspard Monge, for the express purpose
of simplifying the solution of structural problems upon which he
was working for the French government. Therefore an engineer
does not think of lines and planes as being something abstract.
He sees them as something real, concrete, and as a vital part of
the object he is visualizing. His ability to deal with lines and
planes determines his ability to draw correctly all the necessary
views of a structure. Sections 2.12 and 2.17 have shown that a
real practical object may be composed of just lines and planes
and that a logical analysis of these lines and planes may furnish
the very key to the desired solution. Most problems may be analyzed and solved by dealing simply with their elementary component parts, namely, the line and the plane.

The purpose of this chapter is to introduce several line and
plane problems which frequently occur in practice and which all

engineers should know how to solve. These problems may be considered descriptive geometry, but they are just advanced ortho-graphic drawing. Their solution requires knowledge of the relationships between orthographic views and of the method for obtaining any desired view. It also requires the constant use of the seven rules, the fundamental principles, and the four fundamental views as explained in the previous chapters. The same method of logical thinking and three-dimensional visualization will apply, as well as the same notation. In fact, the author is most anxious for the student to realize that the problems to be introduced are all to be solved by the same orthographic methods that were explained in Chapters 1 and 2.

3.2   Theorems

A theorem is a general statement of a truth, which is capable of being proved. The following theorems are given, mostly without proof, for the purpose of familiarizing the reader with certain truths regarding lines and planes. Not only will their study give considerable practice in visualizing lines and planes in space, but the knowledge of the truths which are set forth will greatly aid in clear analysis of the method to use in solving a problem. It is therefore strongly recommended that they be thoroughly studied until they have become self-evident truths.

Theorem 1. If two lines are parallel to each other in space, they will appear parallel (or as points) in all orthographic views.

Theorem 2. If two lines appear parallel to each other in one view, they are not necessarily parallel to each other in space. If two lines appear parallel to each other in two related views, the lines are parallel to each other in space unless they are profile lines. In this special case an extra view is required to determine whether or not the lines are parallel in space. See Figs. 3.1 and 3.2.

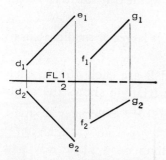

FIG. 3.1. Lines parallel in one view, but not parallel in space.

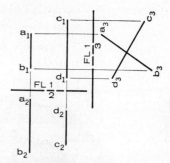

Fig. 3.2. Profile lines parallel in two related views, but not parallel in space.

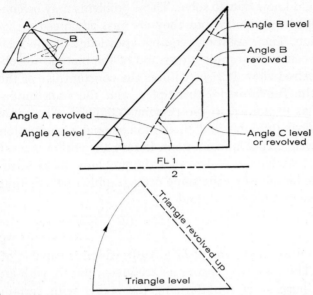

FIG. 3.3. Showing how angles may project.

Theorem 3. If two intersecting lines are at right angles to each other in space, they will appear at right angles to each other in any view which shows the true length of either line. (The one exception occurs when the other line shows as a point.)

This theorem may easily be illustrated by laying a 45° triangle flat on the desk and then revolving it up about one leg, *BC*, which remains on the desk, until it appears as shown by the dashed line in Fig. 3.3. The angle at *C* is the right angle on the triangle, and it still appears as a right angle in the revolved position. In the revolved position one of its legs shows in its true length, but the other leg does not show in its true length. However, the angle still projects as a right angle.

Theorem 4. If two intersecting lines make any angle with each other except a right angle, the view which shows the true size of this angle must show the true length of both lines.

Theorem 5. Any plane angle may project larger or smaller than its true size.

In Fig. 3.3, again, the angles *A* and *B* are each 45°, in their true size. The revolved position of the triangle shows the angle *A* appearing larger than its true size and the angle *B* appearing smaller.

Theorem 6. An infinite number of lines on a plane may pass through the same point on that plane. Only one of these lines can have the same slope as the plane, and that is the one which makes a right angle with a level line on the plane. See Fig. 3.4.

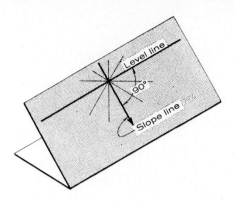

FIG. 3.4. Line having same slope as plane.

Theorem 7. No line on a plane can have a greater slope than the plane itself has. The slopes of all lines lying on a plane vary from zero to a maximum value equal to the slope of the plane.

Theorem 7 can be easily demonstrated by placing a pencil in several positions on a sloping plane and observing the different slopes it can possibly have.

Theorem 8. If two lines lie on the same plane either they must be parallel or they must intersect each other.

Theorem 9. Two lines that lie in two different intersecting planes cannot be parallel to each other unless they are both parallel to the line of intersection of the two planes.

3.3 **To project a point on a line from one view to another view when the line is parallel to a profile plane**

Analysis.   The location of the point on the line would be known in one view, but it would be impossible to project it directly to the other view. Accordingly any new view of the line must be drawn and the point located on the line in this view. This new view may be a side elevation, any auxiliary elevation, or any inclined view. The point, being located in the two views, may easily be located in all other views by projection.

Explanation (see Fig. 3.5). The plan and the front elevation of the line AB are given. The location of the point X on the line AB is known only in the front elevation. It is evident that the point X could not be projected directly to the plan. A new view, 3, of the line AB is drawn. The point X is located on the line AB in this view by the use of Rule 1, as it must be A distance below the folding line. It is then projected directly to the plan from this view. Also any auxiliary elevation could be drawn from the plan, and the point X located in the same way. Also any inclined view

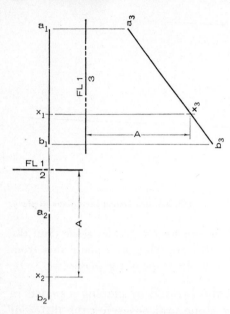

FIG. 3.5. Locating a point on a line in other views.

could be drawn from the front elevation, and the point X located on the line *AB* in this view by projection and again in the plan by measurement according to Rule 2.

3.4   To project a point which is on a given plane, having only one view of the point given

   Analysis.   The given plane would be represented by either parallel or intersecting straight lines. If the given point is on one of the given lines of the plane in one view, it will be on that line in every view; it can be located on that line in any view by simple projection. If the given point is not on any of the given lines of the plane, an auxiliary line is drawn on the plane so that it contains the given point and intersects the given lines of the plane. This auxiliary line is then projected to any other view, and the given point is located on that line in that view.

   Also an edge view of the plane may be drawn and the given point located on that edge view either by elevation or by projection.

3.5   Practice problems

   See Chapter 8, Group 12.

3.6   To draw the plan and front elevation of a line, having given its bearing, its true slope, and its true length

   Note: In problems involving bearing, north is always taken at the top of the paper, and the front elevation is always assumed to be looking due north unless otherwise specified.

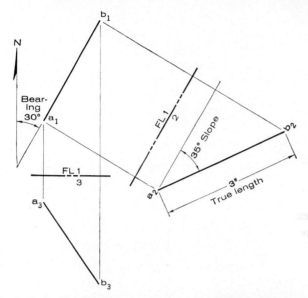

FIG. 3.6. A line having a definite bearing, slope, and length.

Analysis. If the bearing of a line is given, the direction of the plan view is absolutely fixed and the plan view may be drawn indefinite in length. With the plan view fixed, a new elevation with lines of sight at right angles to the line will show its true length and its true slope. In this new elevation view the line may be drawn in, with its true slope and its true length. Both ends of the line are now determined, and may be projected back to the plan or to any other view.

Explanation (see Fig. 3.6). The bearing of the line *AB* is given as N30°E from the point *A*. The true slope is known to be −35° and the true length is known to be 3 in. The plan view of the line *AB* is drawn first, disregarding its length. View 2 is drawn next with lines of sight at right angles to the line *AB*. In this view the line *AB* shows its true slope and its true length as given. This determines both ends of the line, for the point *A* may be placed at any elevation. Both ends of the line are then projected back to the plan and from there to the front elevation, by Rule 1. It is now possible to locate the line in any other view desired.

3.7   Practice problems

See Chapter 8, Group 13.

3.8   To find the perpendicular distance from any point to any line

First method, or line method

Analysis. In the view which shows the given line as a point the required perpendicular distance shows in its true length, be-

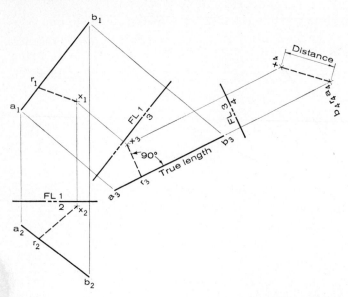

FIG. 3.7. Distance from a point to a line. Line method.

cause it lies at right angles to the lines of sight for this view. The foregoing statement may be easily visualized by holding up a 90° triangle. The view which shows one leg of a 90° angle as a point will also show the other leg of that angle in its true length. The solution, then, consists in simply obtaining a view of the given point and the given line which will show the given line as a point (see Section 2.9).

Explanation (see Fig. 3.7). The line AB and the point X are given in the plan and front elevation views. A new elevation view, 3, is drawn to show the line in its true length. The inclined view 4 is then drawn to show the line AB as a point, and the point X is shown in this same view. The distance from x_4 to a_4b_4 is the true length of the perpendicular distance from the point to the line. It might be required to find where this perpendicular would meet the line AB. Since the perpendicular is at right angles to the line AB in space, it projects at right angles to the line in view 3, because the line shows in its true length in this view (see Theorem 3). The perpendicular is shown as a dashed line in all views.

Note: If the true slope of this shortest connection were required, view 4 would be omitted. Instead, an elevation view off the plan would be drawn, showing the true length and slope of the shortest connection.

Second method, or plane method

Analysis.   A plane may be passed through any three points. The three points A, B, and X therefore fix a plane. The true size of this plane is determined exactly as in Section 2.17. In this true-size view of the plane the perpendicular distance from X to line AB may be drawn, and it will show in its true length. The drawing is not shown here, for it would exactly reproduce the method followed in Fig. 2.15.

3.9　Practice problems

See Chapter 8, Group 14.

3.10　Mathematical method

When more accurate results are desired for this or any other point-line-plane problem, they may be obtained in two ways:

1. By drawing the problem to a much larger scale.
2. By applying simple mathematics to the views of the graphical solution. This rarely requires more mathematics than the solution of right triangles.

As an illustration, the following equations give the complete mathematical solution for a problem quite similar to the problem solved graphically in Fig. 3.7. The notation in the calculation refers to Fig. 3.8.

In views 1 and 3, by the sum of the squares:

$$a_1 b_1 = \sqrt{32^2 + 40^2} = 51.23$$
$$a_3 b_3 = \sqrt{51.23^2 + 24^2} = 56.57$$

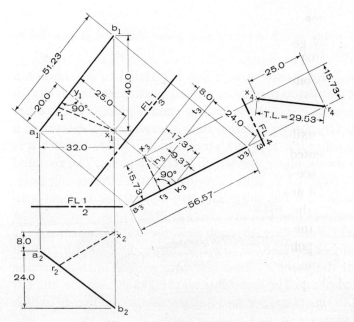

FIG. 3.8. Mathematical solution.

In view 1, using similar triangles $a_1b_1x_1$ and $y_1b_1x_1$:

$$x_1y_1 = \frac{(32)(40)}{51.23} = 25.0$$

In view 1, by the sum of the squares:

$$a_1y_1 = \sqrt{32^2 - 25^2} = 20.0$$

In view 3, using similar triangles $a_3t_3b_3$ and $a_3h_3k_3$:

$$h_3k_3 = \frac{(24)(20)}{51.23} = 9.37$$

In view 3, by addition:

$$x_3k_3 = 9.37 + 8 = 17.37$$

In view 3, using similar triangles $x_3r_3k_3$ and $a_3t_3b_3$:

$$x_3r_3 = \frac{(51.23)(17.37)}{56.57} = 15.73$$

In view 4, by the sum of the squares:

$$\text{True length of } XR = x_4r_4 = \sqrt{25^2 + 15.73^2} = 29.53$$

Notice that the only mathematics required is the sum of the squares of the sides of a triangle and the proportionality of the sides of similar triangles. Simple trigonometry may also be used. This method may be used to obtain calculated results for any point-line-plane problem.

3.11   **To draw a plane which contains one given line and is parallel to another given line**

Analysis.   By geometry, if a line is parallel to any line on a plane, it is parallel to the plane. A third or auxiliary line is drawn which intersects one of the given lines and is parallel to the other given line. The two lines which now intersect must determine a plane. The other line must be parallel to this plane because it is parallel to the auxiliary line which is on the plane. The required plane is represented in both views by the two intersecting lines.

Explanation (see Fig. 3.9). The two nonintersecting and nonparallel lines, AB and CD, are given in both views. Through any definite point on the line AB, such as X, the line EF is drawn so as to be parallel to the line CD in space. Since the lines AB and EF intersect at the point X, these two intersecting lines determine the required plane. The line AB is on this plane, and the line CD is parallel to this plane because it is parallel to the line EF which is on the plane.

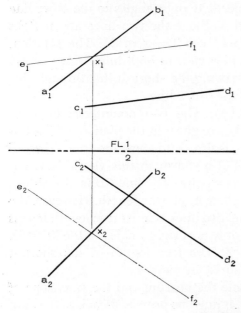

FIG. 3.9. Plane parallel to one line and containing another line.

In order to check this solution, a new view may be drawn showing the plane *ABEF* as an edge and showing the line *CD* in this same view. The line *CD* should check parallel to the edge view of the plane.

A plane could also have been drawn so as to contain the line *CD* and be parallel to the line *AB*.

3.12   Practice problems

See Chapter 8, Group 15.

3.13   To determine the shortest distance between any two nonintersecting, nonparallel lines

First method, or line method

Analysis.   The shortest distance from any point to any line is the perpendicular distance from the point to the line. The shortest distance between two lines will have to be measured perpendicular to both lines. At one, and only one, definitely fixed position in space is it possible to have a line perpendicular to two other lines as specified in this problem. This statement may be visualized by holding up two pencils so that they are nonparallel and nonintersecting, and by trying to see where this shortest distance would be measured. This common perpendicular to both lines appears in its true length in the view which shows either one of the lines as a point. Also in this same view, this perpendicular, being in its true length, projects at right angles to the other line that is not in its true length because the two lines are at right angles to each other in space (see Theorem 3). The solution, then, consists in drawing a new view of both lines which shows one line as a point and in drawing the shortest line through this point and perpendicular to the other line.

Explanation (see Fig. 3.10). The two nonintersecting and nonparallel lines, AB and CD, are given in the plan and front elevation views. A new view, 3, is drawn to show the true length of the line AB. The line CD is also shown in this view. View 4 is drawn to show the line AB as a point. The line CD is also shown in this view, but it does not show in its true length. However, the common perpendicular to the two lines is in its true length in this view and may therefore be drawn at right angles to the line CD. This determines the point Z. The true length of the shortest distance may now be measured in view 4.

If it is desired to determine the bearing and the true slope of this common perpendicular, it must be projected back to the plan view. The point Z is projected to the line CD in view 3, and the perpendicular is then drawn from z_3 at right angles to the line AB, which is in its true length in this view. The intersection determines the point X. Both points, X and Z, may now be projected to any desired view. The bearing of the line XZ is read in the plan. To determine its true slope, a new elevation view, 5, must be drawn showing the line in its true length. The true length

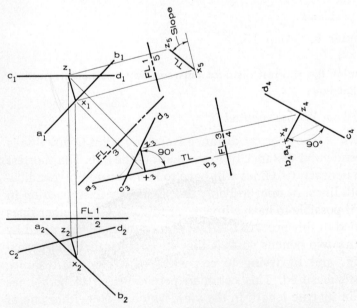

FIG. 3.10. Shortest distance between two lines. Line method.

as found in view 5 should check with the true length found in view 4.

Second method, or plane method

Analysis.   If a plane is drawn containing one of the lines and parallel to the other, an elevation view showing this plane as an edge will show the other line to be parallel to the plane. It will also show the true length and true slope of the shortest or perpendicular distance between the two lines. This is true because this perpendicular distance would have to be perpendicular to the plane assumed, and, since the plane shows as an edge, a line at right angles to the plane will have to show in its true length.

However, the exact location of this perpendicular would not yet be determined. A new view of the line and the plane taken at right angles to the plane will show both the given lines in their true lengths. The common perpendicular to these two lines will then appear as the point where the two lines appear to intersect.

Explanation (see Fig. 3.11). The two nonintersecting nonparallel lines, *AB* and *CD*, are given in the plan and front elevation views. By the method of Section 3.11, the plane *CDF* is so drawn that it contains the line *CD* and is parallel to the line *AB*. View 3 is drawn, showing this plane as an edge and also showing the line *AB* to be parallel to the plane. Another view, 4, is drawn showing the plane in its true size and both the given lines in their true length. In view 4, where the two lines appear to intersect,

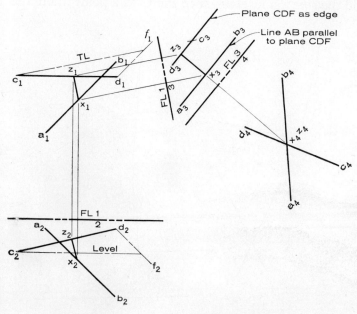

FIG. 3.11. Shortest distance between two lines. Plane method.

the common perpendicular to these two lines appears as a point at x_4z_4. This perpendicular may now be projected back to all other views; its true length and true slope are both shown in view 3. Note that x_1z_1 must always be parallel to FL 1-3.

The engineering application of this problem is made in determining the shortest distance between two tunnels, the location of 90° fittings for connecting two pipes, or the shortest distance between two electric wires.

3.14　Practice problems

See Chapter 8, Groups 16 and 17.

3.15　To determine the shortest level line connecting two nonintersecting, nonparallel lines

Analysis.　A plane is drawn containing one of the lines and parallel to the other line. In an auxiliary elevation view showing the plane as an edge the other line shows parallel to the plane. An infinite number of level lines, all lying at different elevations, could connect these two given lines but they all must lie parallel to the folding line in this auxiliary elevation. The exact elevation at which the shortest one lies is still unknown. Since no level connection can be shorter than the actual level distance between these two lines as it appears in this auxiliary elevation, this apparent distance must be the shortest level distance in its true length. One more elevation, related to the auxiliary elevation, is drawn and the shortest level distance shows as a point where the

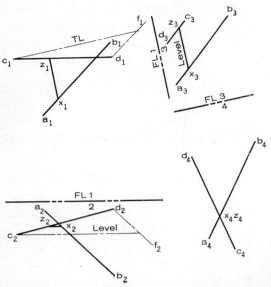

FIG. 3.12. Shortest level distance between two lines.

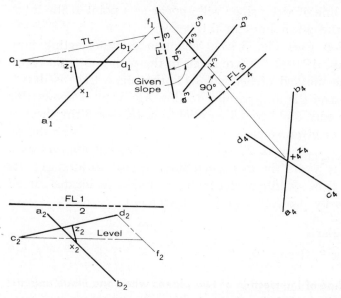

FIG. 3.13. Shortest line of given slope between two lines.

two given lines appear to intersect, because it was in its true length in the preceding view.

Explanation (see Fig. 3.12). The two nonintersecting nonparallel lines, AB and CD, are given in the plan and front elevation. By the method of Section 3.11, the plane CDF is drawn containing the line CD and parallel to the line AB. View 3 shows this plane as an edge and line AB parallel to it. In view 3 the shortest level connection is in its true length and lies parallel to the folding line and in view 4 it appears as a point at x_4z_4 where the two given lines appear to intersect. Its position is now definitely determined and it is located in all other views by projection and measurement. Note again that x_1z_1 is parallel to FL 1-3.

3.16   Practice problems

See Chapter 8, Group 18.

3.17   To determine the shortest line of given slope connecting two nonintersecting, nonparallel lines

Analysis.   As in Sections 3.13 and 3.15, a plane is drawn containing one of the lines and parallel to the other line. In an auxiliary elevation view showing the plane as an edge the other line shows parallel to the plane. In this auxiliary view an infinite number of lines of the given slope could be drawn at different elevations, all parallel to each other, but the exact location of the shortest connecting one is unknown. If a folding line is drawn perpendicular to these lines, and a new view taken, the shortest connecting line of given slope will appear as a point at the intersection of the given lines in this view.

Explanation (see Fig. 3.13). The two nonintersecting non-parallel lines, AB and CD, are given in the plan and front elevation. By the method of Section 3.11, the plane CDF is drawn containing line CD and parallel to line AB. View 3 shows this plane as an edge and line AB parallel to it. In view 3 the shortest connection of given slope is in its true length and makes the required slope angle with FL 1-3, and in view 4 it appears as a point at $x_4 z_4$ where the two given lines appear to intersect. Its position is now definitely determined, and it is located in all other views by projection and measurement.

3.18   Practice problems

See Chapter 8, Group 19.

3.19   To find the line of intersection of two planes when one plane appears as an edge in one of the given views

Analysis. By solid geometry, two planes must intersect in a straight line, every point of which lies on both planes. Also, any two points determine the direction of a straight line. In the view in which one of the planes appears as an edge the points will be apparent where any two lines on the other plane pass through, or pierce, the edge view. Those two piercing points are on both

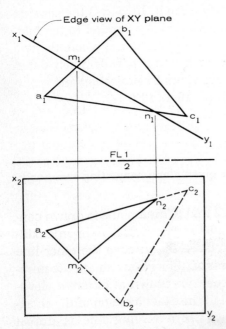

FIG. 3.14. Line of intersection between two planes, one given as an edge.

planes, and therefore they are on the line of intersection of the two. They will be sufficient to determine the line of intersection in all views.

Explanation (see Fig. 3.14). The two given planes are the plane ABC and the plane XY, which shows as an edge in the plan view. The lines AB and AC are both on the plane ABC. In the plan view it is apparent that the lines AB and AC pass through the plane XY at the points M and N, respectively. Therefore the points M and N must lie on both the given planes and therefore on the line of intersection of the two. In any other view than the plan, M and N are located by projection, and in all views they determine the direction of the required line of intersection, regardless of its length. The portion AMN of the plane ABC is in front of, and the portion MNCB is behind, the plane XY.

3.20   **To find where a line pierces an oblique plane**

Analysis.   A new view is drawn showing the plane as an edge. The line is shown in this same view. The exact point at which the line pierces the plane will be apparent in this view, just as it was in Section 3.19. The new view to be drawn may be either an elevation edge view or an inclined edge view.

3.21   **To find where a line pierces an oblique plane, using only the two given views**

First method: by vertical projecting plane

Analysis.   The vertical projecting plane for the given line is the plane which contains all the vertical lines of sight for the line, or it is the plane which actually projects the line onto the plan image plane. This vertical projecting plane shows as an edge in the plan view, and, of course, contains the given line. By the method of Section 3.19, the intersection of this vertical projecting plane with the given plane is determined. The given line and this line of intersection both lie in the vertical projecting plane, and therefore they must either intersect or be parallel. This relationship cannot be seen in the plan because the entire plane appears as an edge. However, it is apparent in the other given view. If the two lines intersect, the point of intersection is on both the given line and the given plane and is therefore the point at which the given line pierces the given plane.

Explanation (see Fig. 3.15). The line XZ and the oblique plane ABC are given. The vertical projecting plane of the line XZ appears as an edge in the plan as x_1z_1. The intersection of this

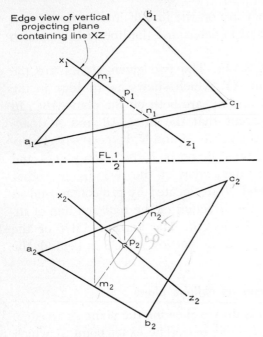

Edge view of vertical
projecting plane
containing line XZ

FIG. 3.15. Line piercing an oblique plane. Vertical-projecting-plane method.

vertical plane with the plane ABC is found to be the line MN, which is shown in both views. The lines MN and XZ both lie in the vertical projecting plane, and in the front elevation these lines are seen to intersect at the point P. This point is on the line MN, every point of which is on the plane ABC, and therefore it is also on the plane ABC. But it is also on the line XZ. Therefore it is the required point where the line XZ pierces the plane ABC. If the front elevation had shown the line XZ to be parallel to the line MN, that would have been proof that the line XZ was parallel to the plane, because it was parallel to a line MN on the plane. It is evident that there could be no piercing point in that case.

Second method: by front projecting plane

Analysis.   The front projecting plane of the line is the plane which projects the line upon the front image plane. The line of intersection of the front projecting plane with the given plane is determined, and the method from here is exactly the same as the first method in Section 3.21. However, in this case the piercing point is apparent in the plan view instead of in the front elevation.

Explanation (see Fig. 3.16). The line XZ and the oblique plane ABC are given. The front projecting plane containing the line XZ appears as an edge at x_2z_2. The intersection of this plane

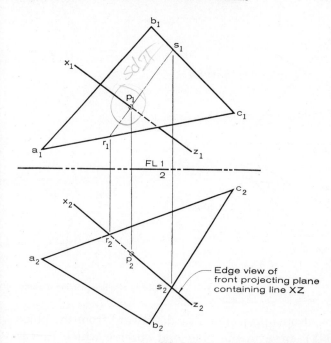

FIG. 3.16. Line piercing an oblique plane. Front-projecting-plane method.

with the plane ABC is found to be the line RS. In the plan it is apparent that the lines XZ and RS intersect at the point P. This point is therefore on both the given line and the given plane, and is the point required.

The methods of Sections 3.20 and 3.21 furnish three absolutely independent ways for determining where a line pierces an oblique plane. All three methods should determine exactly the same point, and if they are all used on the same problem they will furnish an excellent check on the accuracy of the drafting work. In most engineering settings it is easier to solve the problem in just the two views given. But in some cases it may be easier to draw the new edge view. Judgment must always be used by the draftsman in selecting the best method to use for each job.

3.22   Practice problems

See Chapter 8, Group 20.

3.23   To determine the visibility of lines

When two lines cross but do not intersect, it is important to determine which line is visible and which line is obscured by the other. Looking down on the plan view, whichever line is uppermost will be visible. Looking in horizontally at the elevation view, whichever line is nearer the observer will be visible.

In Fig. 3.17, to determine whether line BC or XY is uppermost

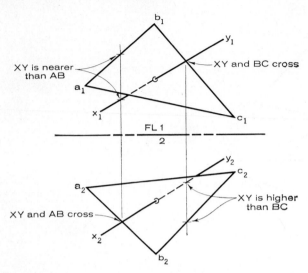

FIG. 3.17. Visibility of a line piercing a plane.

in the plan view, drop a projection line downward from the place where BC and XY cross in the plan and observe which line is higher in the elevation view. XY is higher than BC and therefore XY is visible at this point in the plan view and continues to be until it passes through the piercing point when it is underneath the plane ABC and invisible. Check the crossing of XY and AC in the plan view to confirm this. To determine visibility in the elevation view, draw a projection line upward from the crossing of XY and AB in the elevation view and observe which line is nearer the observer in the plan view. XY is nearer than AB and therefore XY is visible at this point in the elevation view and continues to be until it passes through the piercing point where it becomes invisible. Check the crossing of XY and AC in the elevation view to confirm this.

This method is a general one and can be applied not only to piercing point problems, but to all situations where visibility is in question.

3.24 To find the line of intersection of any two oblique planes

Edge-view method

Analysis.   A new view, elevation or inclined, may be drawn showing one of the planes as an edge, and including the other plane. This new view will indicate the points at which any two lines on the second plane pierce the plane which appears as an edge. The line of intersection is determined by these two piercing points.

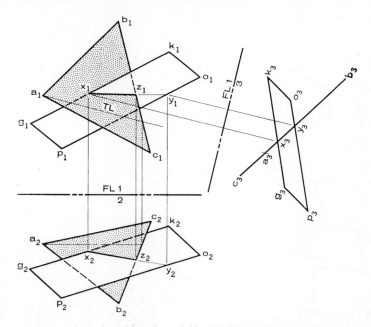

FIG. 3.18. Line of intersection of two planes. Edge-view method.

Explanation (see Fig. 3.18). The two oblique planes ABC and $GKOP$ are given. An auxiliary view is drawn in which ABC appears as an edge. In this view line GK of plane $GKOP$ pierces plane ABC at point X, and line OP pierces plane ABC at Y. Points X and Y are projected back to the plan and elevation views of GK and OP, respectively, and XY is the line of intersection between the two planes. Notice that, although the plan and elevation views show that Y lies beyond the boundary of plane ABC, Y is nevertheless on plane ABC if its limits were extended. The part of XY that is common to the two given planes is XZ and therefore XZ is the actual line of intersection of the given bounded planes. The vertical alignment of the plan and elevation views of Z is a check on the accuracy of the solution.

3.25 **To find the line of intersection of any two oblique planes, using only the two given views**

First method: line method

Analysis.   Any line on one of the planes may be selected, and the point at which it pierces the other plane may be determined by the methods of Section 3.21. This point lies on both planes. The same process may be repeated for some different line, which will determine a second point on both planes. These two points determine the entire line of intersection of the two given planes.

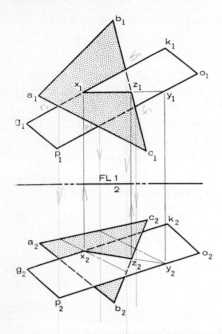

FIG. 3.19. Line of intersection of two planes. Line method.

However, if two more points were to be determined in the same way, all four points should prove to be in a straight line.

Explanation (see Fig. 3.19). The two oblique planes ABC and $GKOP$ are given. By the method of Section 3.21, GK is found to pierce ABC at point X and OP pierces ABC at Y. XY is the line of intersection of the two planes. Notice that, although Y lies beyond the boundary of plane ABC, Y is nevertheless on plane ABC if its limits were extended. The part of XY that is common to the two given planes is XZ and therefore XZ is the actual line of intersection of the given bounded planes. The vertical alignment of the plan and elevation views of Z is a check on the accuracy of the solution.

Second method: auxiliary-plane method

Analysis. A third or auxiliary plane, sometimes called a cutting plane, may be drawn in any position in either view, if it appears in one view as an edge. The intersection of this plane with each of the given planes may be found by the method of Section 3.19. These two lines of intersection both lie on the auxiliary plane, and therefore they must either intersect or be parallel. In case they intersect, the common point on the two lines of intersection will be apparent in one view and may be projected to the other. This point is on both the given planes, and therefore is on their line of intersection. The process is repeated by taking another auxiliary plane, one point on the line of intersection being determined by each plane. Two points thus determined are sufficient to establish the required line of intersection, although a third point should always be determined for a check.

Explanation (see Fig. 3.20). The planes ABC and XYZ are the two given oblique planes. The first auxiliary plane to be assumed is a vertical plane which is found to intersect the given planes in the lines HK and RS as shown. The front elevation shows these two lines intersecting at the point M, which is projected back to the plan. This point is on both the given planes, or the planes extended, and is therefore on their line of intersection.

A second auxiliary plane is taken appearing as an edge in the front elevation, and similarly the point N is found to lie on both the given planes. The points M and N are sufficient to determine the direction of the entire line of intersection in any view. It is safer to determine at least three points on the line of intersection, and to see whether they check in a straight line in all views. Any other cutting plane may be drawn at any angle in either view, as long as it appears as an edge in one of the given views. Attention has already been called to the fact that, for purposes of solution, planes may be considered to be indefinite in extent. In this case the points M and N are not inside the limits of the planes as given, but they are on both planes extended. The line of intersection of

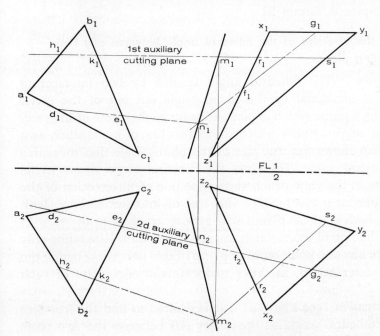

FIG. 3.20. Intersection of two planes. Auxiliary-plane method.

these two planes cannot possibly be in any other position, regardless of how large the planes are.

If the lines HK and RS should happen to be parallel instead of
intersecting, this fact would prove these two lines were parallel
to the line of intersection of the given planes (see Theorem 9).
While the direction of the required line of intersection would
then be known, its position would still be unknown. In such
a case the vertical plane should be assumed at some different angle with the front image plane, and the problem could be
completed. If this different vertical plane still intersects the two
given planes in two parallel lines, the given planes themselves
must be parallel.

When the lines cut out by an auxiliary cutting plane are nearly
parallel and have a broad intersection at a very large or very
small angle, this method is subject to inaccuracy. Care should be
taken to select cutting planes that will produce lines giving good
intersections.

Three independent methods of finding the line of intersection
of two planes have been explained in Sections 3.24 and 3.25.
Each method has its advantages and is more easily applied to
some problems than the other two. The draftsman should be
familiar with all three methods.

3.26   Practice problems

See Chapter 8, Group 21.

3.27   To find the true size of the dihedral angle between any two intersecting planes

Analysis.   The dihedral angle formed by two intersecting
planes is measured by the plane angle cut out of the given
planes by a plane which is perpendicular to their line of intersection. An auxiliary view which shows this line of intersection as a
point also shows the true size of the plane angle that measures
the dihedral angle. Therefore the dihedral angle shows in its
true size in the view which shows the line of intersection of the
two planes as a point. Since this line of intersection is on both
planes, both of these planes will appear as edges in the view in
which the line of intersection appears as a point. It remains only
to locate any one point on each plane in this new view, beside the
line of intersection, in order to locate the edge view of each
plane.

Explanation (see Fig. 3.21). It is desired to find the true size
of the dihedral angle at the valley AB between the two roofs
ABED and ABC, shown in the plan and front elevation. A new

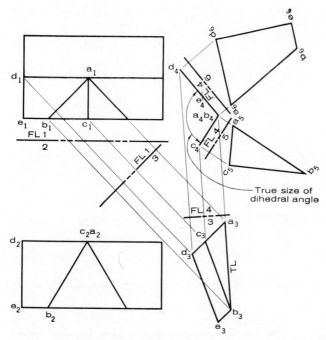

FIG. 3.21. Angle between two planes.

elevation view, 3, is first drawn showing the line of intersection *AB* in its true length. The points *C* and *D* are chosen at random, one on each roof, and they are located in view 3 also. View 4 is drawn showing the line *AB* as a point, and showing also the points *C* and *D*. Since both planes must appear as edges in view 4, these planes may both be drawn in to the line of intersection as a point, since one other point is located on each plane. The dihedral angle is now shown in its true size in view 4, where it may be scaled.

To show how practicable this method is, two more views, 5 and 6, are drawn from view 4, at right angles to each plane. Each plane will then appear in its true size. If this valley angle is to be detailed as a steel-connection angle, views 4, 5, and 6 are all that are required to make a complete shop drawing for steel fabrication. The scale of the drawing, from view 4 on, might easily be enlarged, but the angles, cuts, and lengths all appear in their true size in the three views shown, and the relationship between these views is the same as for any simple shop drawing. Structural-steel draftsmen utilize this method constantly in determining the angles and cuts for bent-plate connections.

In the problem of Fig. 3.21 the line of intersection between the two planes was given. In any other case, it is necessary only to determine this line of intersection as in Sections 3.24 and 3.25, and to proceed from there as in the problem above.

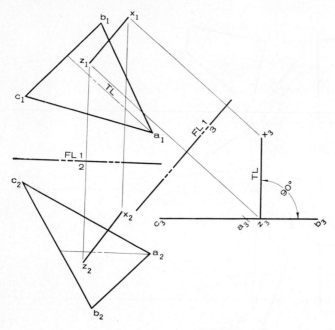

FIG. 3.22. Line perpendicular to a plane. Edge-view method.

3.28   Practice problems

See Chapter 8, Group 22.

3.29   To draw a line perpendicular to a plane from a point not on the plane

Analysis.   A new view of the point and the plane should be drawn which will show the plane as an edge. In this view the perpendicular line from the point to the plane can be drawn at right angles to the edge view of the plane. This perpendicular line appears in its true length in this new view, which information determines the plan view of the line; it is parallel to the folding line or at right angles to the lines of sight for the new view. The point where the perpendicular pierces the plane is also evident in the new view.

Explanation (see Fig. 3.22). The plane ABC and the point X are given in plan and front elevation. View 3 is drawn to show the plane as an edge and to show the point X. From x_3 the perpendicular line may be drawn at right angles to the plane, and it is seen to pierce the plane at z_3. The line XZ shows in its true length in view 3, and therefore the lines of sight for this view must be at right angles to the line XZ in the plan. This fixes the plan view at x_1z_1. The front elevation of the line XZ is obtained by projection.

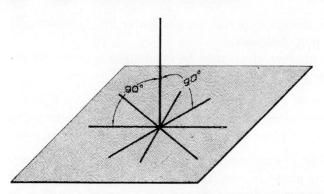

FIG. 3.23. Line perpendicular to a plane and lines in the plane.

3.30   To draw a line perpendicular to a plane from a point not on the plane, using two views only

    Analysis.   By solid geometry, if a line is perpendicular to a plane, it is perpendicular to all lines on the plane that intersect the perpendicular (see Fig. 3.23). By Theorem 3, which states that perpendicularity is apparent in any view when only one of two perpendicular lines is shown true length, if one of these intersecting lines on the plane should show in its true length in some view, the perpendicular to the plane would project at right angles to that line in that view.

    Any number of lines on an oblique plane may show in their true lengths in the two given views, but they are all parallel (see Fig. 3.24). Then any one of these true-length lines on a plane determines the *direction* of the perpendicular to the plane in

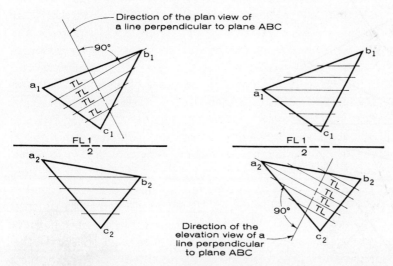

FIG. 3.24. True-length lines establishing the direction of a line perpendicular to a plane.

that view even though it is not the particular line that actually intersects the perpendicular.

The statements just given may be condensed into the following simple rule:

Rule 8. Direction Principle. In any orthographic view, a line that is perpendicular to an oblique plane in space shows at right angles to any line on the plane that shows in its true length in that view.

Caution: Rule 8 determines only the direction of the perpendicular to a plane in any view; it does not determine the point where the perpendicular pierces the plane or establish its length. This piercing point is determined separately by any of the three methods already explained. It must also be remembered that the direction of a line must be found in *two views* before the position of the line itself is fixed in space.

It is permissible to consider two lines as perpendicular to each other even when they do not intersect. To illustrate noninter-secting perpendicularity, two pencils may be held at right angles, first intersecting, and then drawn apart with the perpendicularity maintained. This concept does not alter any principles or rules for the solution of problems.

Explanation (see Fig. 3.25). The plane ABC and the point X are given in the two views. Any level line at random, such as the line AM, is chosen on the plane and is drawn in both views. The perpendicular to the plane from the point X is next drawn in the

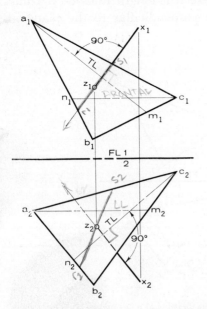

FIG. 3.25. Line perpendicular to a plane. Direction principle.

plan view at right angles to the line *AM*, which shows in its true length in the plan. Since all level lines on the plane are parallel, any other level line on the plane would have given the same direction to the plan view of the perpendicular. This perpendicular to the plane is drawn indefinite in length because the piercing point is still unknown.

In the same way a random frontal line, such as *CN*, is drawn on the plane and the perpendicular to the plane is drawn from the point X at right angles to the line *CN* in the front elevation and indefinite in length. The perpendicular to the plane is now fixed in two views; the point where it pierces the plane is determined by the method of Section 3.21 and is found to be the point Z. A new elevation view is required if it is desired to show the true length and true slope of this perpendicular distance.

3.31   Practice problems

See Chapter 8, Group 23.

3.32   To draw a plane perpendicular to a line and through a fixed point, using two views only

Analysis.   This problem is just the reverse of the problem in Fig. 3.25. In this case the line and only one point on the plane are given, and the shape and size of the plane are both indefinite. Through that one known point on the plane a level line can be drawn in the elevation view. This line will be in its true length in the plan, and therefore, by the direction principle, it will have to show at right angles to the plan view of the given line. In the same way a frontal line can be drawn through the point in the plan view, and its direction will be known in the front view. These two lines, since they intersect, determine the required plane which may then be given any definite shape or size desired.

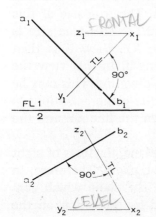

Explanation (see Fig. 3.26). Line *AB* and point X are given. A level line *XY* of any length is drawn through X with its plan view shown true length and at right angles to a_1b_1. A frontal line *XZ* of any length is drawn through X with its front view shown true length and at right angles to a_2b_2. Plane *XYZ* is the required plane perpendicular to *AB* and containing the given point X.

FIG. 3.26. Plane perpendicular to a line through a given point.

3.33   Practice problems

See Chapter 8, Group 24.

3.34   To project a line upon an oblique plane

Analysis.   To project a line upon a plane means to draw a series of lines from the given line *perpendicular to the given plane* and to find where these perpendiculars pierce the plane. Two perpendiculars, one from each end of the line, are sufficient to project a straight line upon a plane. The method of drawing the perpendiculars and finding their piercing points is exactly the same as in Section 3.30. The line connecting the two piercing points is the projection of the line upon the plane.

3.35   Practice problems

See Chapter 8, Group 25.

3.36   To find the true size of the angle a line makes with a plane

First method

Analysis.   The angle a line makes with a plane is the angle between the line and its projection on the plane. This angle lies in a projecting plane containing the line and perpendicular to the given plane. In order to show this angle in its true size the *edge view of the plane* and the *true length of the line* must be seen in the same view. First, a view should be drawn showing the plane in its true size and also showing the line. Any view related to this one shows the plane as an edge. A view is then drawn from the view showing the true size of the plane, to show the true length of the line. The desired angle between the line and the plane is seen in this view.

Explanation (see Fig. 3.27). The plane *ABCD* and the line *XY* are given in plan and front elevation. The true size of this given plane is first found as in Section 2.17. This true size is shown in view 4, together with the line *XY*. View 5 is then drawn to show the line *XY* in its true length. In this view the plane appears as an edge parallel to the folding line. This may be checked by measuring several points on the plane according to Rule 2. The true size of the angle between the line *XY* and the plane may be scaled in view 5.

Since view 4 shows the true size of the plane, the lines of sight for this view must be perpendicular to the plane. Therefore, in view 4, the line x_4y_4 is the edge view of the plane which contains the angle between the line *XY* and its projection on the plane.

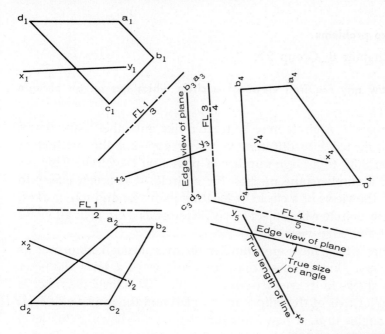

FIG. 3.27. Angle between a line and a plane.

In solving problems by this method it should always be remembered that *the first thing to do is to find the true size of the plane.* This true-size view is the *key view*.

Second method

Analysis.   It has already been stated that the angle a line makes with a plane is the angle between the line and its projection on the plane. The line and its projection on the given plane both lie in the projecting plane whose true size may be found by the ordinary methods. This method will give the true size of the angle required with one less view than the first method, but it is harder.

3.37   Practice problems

See Chapter 8, Group 26.

3.38   To draw any required views of a plane figure which lies on an oblique plane

Analysis.   A view should be obtained first which shows the true size of the plane. The given plane figure may then be drawn on the plane in its true size, being so placed on the plane as to satisfy whatever conditions were specified. This plane figure may then be projected, point by point, back to all other views, by the ordinary methods of orthographic projection.

3.39   Practice problems

See Chapter 8, Group 27.

3.40   To draw any required views of a circle which lies on an oblique plane

Analysis. A circle may be treated like any other plane figure and projected into different views by applying the method of Section 3.38 to a large number of points on the circle. This becomes a cumbersome method for a circle. It is much easier to obtain the views of a circle by locating its major and minor axes, because a circle will project as an ellipse or as a circle in all views except the edge view.

A circle may have any number of lines through its center as diameters. One of these diameters is always a level diameter which shows in its true length in the plan. This level diameter is the major axis of the ellipse in the plan and the minor axis is at right angles to it. A new view may be drawn showing this level diameter as a point; the plane of the circle is an edge in this view. On this edge view the circle may be laid off in its true length, or diameter, and projected back to the plan view to give the length of the minor axis. The two axes having been determined, the ellipse may be drawn in by the card or trammel method (see Appendix A.2) to complete the plan view.

In exactly the same way the front elevation may be obtained by choosing a diameter of the circle which is a frontal line and proceeding as before. Any other view may be obtained as easily by choosing a diameter of the circle that shows in its true length in that view.

Explanation (see Fig. 3.28). The plane *ABC* and the point *X* on this plane are given. The point *X* is the center of a circle having a given diameter and lying on the plane. In the plan view a level diameter is drawn through the point *X* parallel to any level line on the plane, such as the level line through the point *A*. This level diameter is laid off in its true length, the actual diameter of the circle, to give the major axis for the ellipse in the plan. View 3 is drawn to show the plane as an edge. The major axis for the plan appears as a point in this view at x_3, which is also the center of the circle. In view 3 the circle is laid out in its full diameter and then projected back to the plan to give the length of the minor axis. The ellipse is drawn in by the trammel method (see Appendix A.2).

To obtain the front elevation a frontal diameter is drawn through the point *X*, parallel to a frontal line on the plane, and is

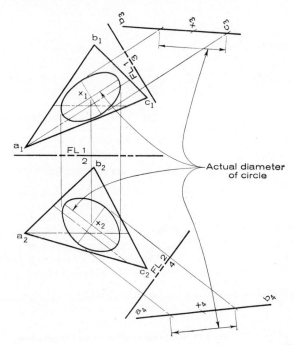

FIG. 3.28. A circle lying on an oblique plane.

laid off in its true length to give the major axis for the front elevation. The new view 4 is drawn from the front elevation as shown
and the procedure is exactly the same as for the plan view.

The method which has just been explained will be found to be
the easiest, fastest, and most accurate way of drawing circles on
inclined planes. It should be remembered that the major and
minor axis for each view are entirely different sets of lines. When
the related views of a circle have been completed, each point of
the circle should project perfectly between views. The extreme
projecting points in related views furnish a quick check.

3.41   Practice problems

See Chapter 8, Group 28.

3.42   To draw views of a solid object resting on an oblique plane

Analysis.   The solid object can usually be drawn in its simplest
position in the view showing the plane as an edge. Whether it
can or not depends upon the object itself. In some cases a view
might have to be drawn showing the true size of the plane before
the object could be drawn in its correct position. After the object
has been placed in the correct position in either the edge or the
true-size view of the plane, it may easily be projected to all other
views.

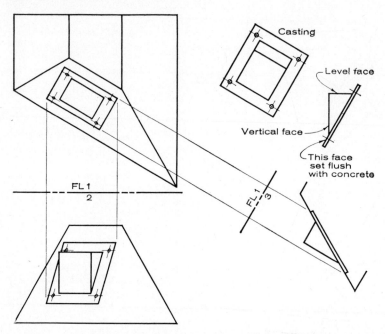

FIG. 3.29. Solid object resting on an oblique plane.

Explanation (see Fig. 3.29). The plan and front elevations of the end portion of a concrete pier are given. Also two views are given of a casting which is to be placed in the end face of the pier, as specified on the drawing. The new view 3 is first drawn showing the end face of the pier as an edge. The casting is then drawn in position in this view with its proper face set flush with the concrete as specified. The exact location of the casting on the face was not specified in this problem. A view could be drawn off from view 3 to show the other view of the casting in position. However, in this case such a view is not necessary, for it would be merely a reproduction of one of the given views of the casting. Any necessary measurements may be taken from the given view, if it is drawn to the same scale or from the given dimensions.

3.43 Practice problems

See Chapter 8, Group 29.

3.44 Mining problems

The principles which have been explained thus far are quite applicable in the solution of problems encountered by the mining engineer. A vein of ore or a stratum (layer) of rock, within certain limits, may be considered to be just a plane having some thickness and nearly always some incline. The terms commonly

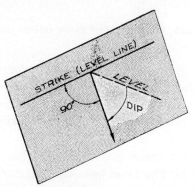

FIG. 3.30. Pictorial illustration of strike and dip.

FIG. 3.31. Plan or map-view method of indicating strike and dip.

used by mining engineers for giving the lay of a vein of ore are the "strike" and "dip."

The *strike* of a vein is the bearing of a level line on the plane of the vein.

The *dip* of a vein is the true slope of the vein, and is the angle measured *downward* from a horizontal plane and at right angles to the strike. Figure 3.30 shows pictorially the meaning of strike and dip.

When measuring strike and dip in the field the geologist locates a visible exposure called an "outcrop" of a vein, applies an instrument operating on the principle of a spirit level and compass called a "clinometer" to the exposed upper or lower plane of the vein, and records the strike and dip in their exact location on a map as shown in Fig. 3.31. The dip angle is always downward in the direction of the arrow. There is only one plane that can be fitted into the T-shaped figure formed by the strike and dip and therefore the position of the plane is definitely fixed by this device.

When referring to dip verbally when the map symbol is not given, it is important to designate its general direction. For example, referring to Fig. 3.31, a plane dipping 43° NW is quite different from one having the same strike and dipping 43° SE.

If it is impossible to measure strike and dip from a single outcrop because of weathering or other physical reasons, the vein may be located by using the outcrop as one point in combination with other points located by sinking boreholes until the vein is met. A total of three definite points on the same side of the vein are required. A vein has some thickness and is therefore bounded by two parallel planes, either one of which may be used to locate it. Any problem in relation to this vein may be solved by treating the three points on the vein as a plane and by using the methods previously described for the solution of planes.

3.45   To determine the strike of a vein of ore, having given three points on the vein

Analysis. The three points, as given in the plan and in some elevation, determine a plane in those two views. A level line is drawn on this plane in the elevation view and projected to the plan, where its bearing may be read. The bearing of this level line is the strike of the vein.

Explanation (see Fig. 3.32). The three given points which determine the vein are A, M, and N. The level line AC is drawn on the vein in the front elevation at a_2c_2 and projected to the plan at a_1c_1 where its bearing is found to be N45°W. This is the strike of the vein.

3.46   To determine the dip of a vein of ore, having given three points on the vein

Analysis. The three points, as given in the plan and in some elevation, determine a plane in those two views. Since the dip angle is the same as the true slope angle of the plane, an auxiliary elevation showing the plane as an edge shows its dip. This view is drawn looking parallel to the strike.

Explanation (see Fig. 3.32). A contour map of a small portion of ground is shown. The points M and N are points of outcrop on the upper surface of the vein. A vertical borehole

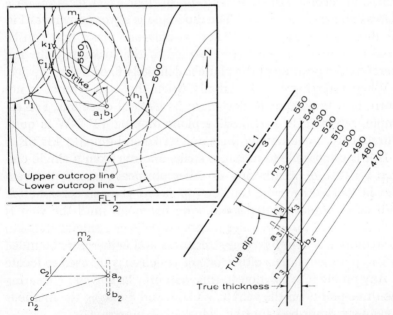

FIG. 3.32. Strike, dip, and line of outcrop of a vein.

intersects the upper surface of the vein at A and the lower surface at B. The three points, A, M, and N, establish the vein in both the plan and front elevation views. The strike is determined as in Section 3.45. The elevation view 3 shows the vein as an edge and also shows its true dip, which is 34°. In this same view the point B on the lower surface is located also and the lower surface drawn parallel to the upper one. The distance between the two planes is the thickness of the vein and it may be measured in this view.

3.47   To determine the line of outcrop of a vein

Analysis.   The line of outcrop is the line of intersection of the plane of the vein with the irregular ground surface. Each point on this line may be found by intersecting a level line on the vein with a contour line on the ground at the same elevation. Several points must be found to determine the irregular line of outcrop.

Explanation (see Fig. 3.32). Referring to the upper plane AMN of the vein, the intersection of the 510-ft contour with this plane as seen in view 3 gives the points h_3 and k_3 which are projected to the plan at h_1 and k_1. These two points are on the vein and also on the ground and are, therefore, two points on the line of outcrop which must be seen in the plan view. Other points determined in the same way at different elevations give the upper outcrop line. In the same manner the lower outcrop line is found. The extent of the vein is thus definitely shown within the limits of the map. Notice that the ore all lies between the arrowheads shown on the map. There is no ore outside of the lower outcrop line.

3.48   To determine the strike, dip, and thickness of a vein by two noncoplanar boreholes

Analysis.   The two noncoplanar boreholes would intersect both the upper and lower surfaces, giving two points on each surface of the vein. Connect the two points on the upper surface with a straight line. A view showing this line as a point shows the upper surface as an edge, although its direction is still unknown. Since the two surfaces are parallel, the lower surface must also be an edge in this view and its position is determined by locating the two points on this surface. The upper surface may now be drawn parallel to the lower one and the thickness is apparent here. Either surface may be projected back to the plan to determine the strike and dip by the usual methods.

Explanation (see Fig. 3.33). A vertical borehole at A intersects the upper surface at B and the lower surface at C. An in-

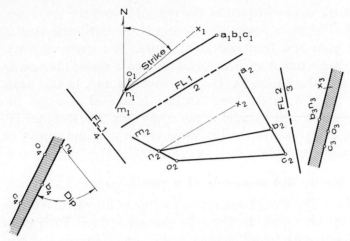

FIG. 3.33. Strike, dip, and thickness by two noncoplanar boreholes.

clined borehole at M intersects the upper surface at N and the lower surface at O. The line BN is therefore on the upper surface and it appears as a point in view 3. Points O and C on the lower surface are located in view 3, where they determine the edge view of the lower surface. The upper surface is drawn parallel to the lower surface and the true thickness is seen. To find the strike draw a level line NX on the upper surface in view 2 and project it to view 3. This level line is now fixed in space and may be projected to n_1x_1 in the plan where the strike is measured. A new elevation, view 4, shows the dip. The true thickness may be seen again in this view if the lower surface is located here. If the two boreholes had been coplanar, the two lines on the upper and lower vein surfaces would have to be parallel and they both would appear as points in view 3. Therefore the position of the vein could not be determined from the data as given.

Alternate method

The line BN is entirely on the upper surface and the line CO is entirely on the lower surface of the vein. By the method of Section 3.11 pass a plane containing one of these lines and parallel to the other line. Draw a new view off the plan showing the plane as an edge and the other line parallel to it. In this new elevation view the upper and lower surfaces of the vein are both edge views and parallel and therefore the dip and thickness of the vein may be measured here. The bearing of any level line on either surface gives the strike of the vein. This method is simpler and requires fewer views than the first method. The solution is not shown.

3.49   Faults

Veins of ore do not always continue indefinitely as plane surfaces (with thickness), as has been assumed in the preceding
sections. Sometimes a dislocation of the structure takes place
which breaks its continuity. Part of a vein may be upthrown or
downthrown by either a translating or a rotary motion. The plane
in which the separation takes place is called the *fault plane*.

The following definitions applying to faults are in use by the
U.S. Geological Survey:

1. *Slickensides* are smooth surfaces on the fault plane having
lineations or grooves which are caused by the motion in separat-
ing and which are evidence of the direction of motion.

2. *Rake* is the angle on the fault plane between the slickensides lineations and a level line on the fault plane.

3. *Net slip* is the true length of the actual distance moved.

4. *Plunge* is the true slope of the net slip (measured in a
vertical plane).

5. *Strike* of the net slip is the bearing or map direction of
the motion.

These definitions are all illustrated in Fig. 3.34. The strike lines
of the original vein, the displaced vein, and the fault plane are
usually taken at the same elevation in solving problems.

3.50   To find the net slip and plunge, having given the original and displaced strikes and dips of the vein, the strike and dip of the fault plane, and the rake

Analysis.   See Fig. 3.34. Since the strike and dip of the fault
plane are known, the edge view (2) and true-size view (3) of the
fault plane can be drawn. In this true-size view the given rake
can be laid off. One of the lineations would be a_3x_3, and this is
projected back to the plan at a_1x_1. Edge view (4) of the original
and displaced positions of the vein may now be drawn because
the dip of each is known. The lineation line AX, which was
chosen in view 3, is now located in view 4 as shown at a_4x_4. This
line now shows in this view not only the direction of the motion
but also the amount of the motion, since the point A on the original vein could only move to B on the displaced vein. This line AB,
which is the amount of the motion, is projected back to the plan
and then to view 5 where the net slip and plunge are found. The
plan gives the bearing or map direction of the motion or the strike
of the net slip.

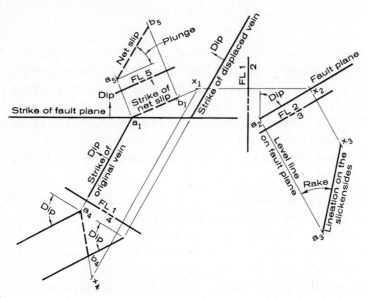

FIG. 3.34. Fault problem. Illustration of definitions.

3.51   To find the apparent dip, having given the true dip, by the geologist's compressed method

Students who have studied descriptive geometry are sometimes confused when first confronted with compressed or short-cut methods used by structural geologists. A description here of these methods may be helpful.

One of the most common problems in geology is to find the apparent dip of a bed of rock when the true dip is given. *True dip* is the slope of a plane at right angles to the strike, and *apparent dip* is the slope of a plane at some angle other than at right angles to the strike.

Analysis.   In Fig. 3.35a line $C'F'$ may be considered another strike parallel to the given strike AD and a known distance BC' or EF' (BC' equals EF') lower in elevation. BE is the plan view of $C'F'$. If the two right triangles ABC' and DEF' are rotated upward about AB and DE, respectively, until they are horizontal, then the plan view will show the true dip and the apparent dip.

Explanation (see Fig. 3.35b). The strike AD and true dip are given in the usual manner. AB is drawn at right angles to the given strike and the true dip angle is laid off at BAC. Line BE is drawn parallel to the given strike at any desired distance from it and BE is extended until it intersects the line making the true dip angle at C. The direction DE of the apparent dip is laid off according to its bearing at any convenient distance from AB.

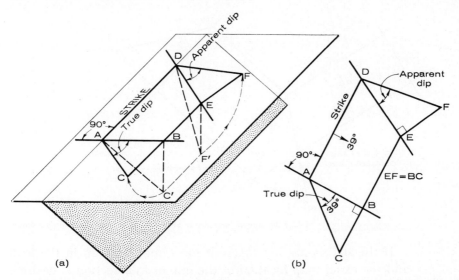

FIG. 3.35. (a) Geologist's method. Pictorial illustration. (b) Geologist's method for finding apparent dip.

EF is drawn at right angles to *DE* and is made equal in length to *BC*. Angle *EDF* is the required apparent dip.

If the strike and apparent dip are given and it is required to find the true dip, the problem can be worked backward laying off the apparent dip first.

One may think of *AC* in Fig. 3.35*b* as an edge view of a plane in an elevation view looking in endwise on a level line (the strike). Notice also that, if the true dip and apparent dip triangles were separated from the strike and a detached folding line inserted between, the problem would then appear in its conventional form.

3.52   Practice problems

See Chapter 8, Group 30.

3.53   To find the bearing and slope of the line of intersection of two beds of rock, having given their strike and dip, by the geologist's compressed method

Analysis.   For convenience in solving this problem the two strikes are assumed to be at the same elevation. The two strikes are level lines at the same elevation, and if they are extended until they intersect, one point on the line of intersection between the two planes can be located. A second strike is located for each of the planes, each the same distance below the given strikes. These new strikes are level lines at the same elevation, and if they are prolonged until they intersect, a second point on the line of intersection is located and the line of intersection fixed.

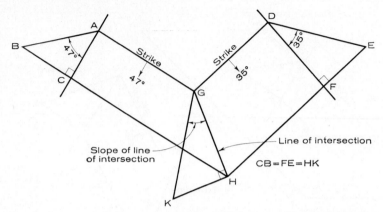

FIG. 3.36. Line of intersection of two beds of rock. Geologist's method.

If the strikes are not taken at the same elevation in the field and the exact map location of the line of intersection is desired, the same reasoning can be used but the problem will not be dealt with here for the sake of brevity. We are concerned only with the bearing and slope, not the exact location, of the line of intersection.

Explanation (see Fig. 3.36). The given strikes are drawn in any convenient place on the paper (their relative location will not affect the answer), and are then prolonged until they intersect at G. Line AC is drawn at right angles to AG and the given dip laid off at BAC. CH is drawn parallel to AG at any convenient distance from AG and is then prolonged until it intersects the dip angle at B. CB is the vertical distance that the new strike CH lies below the given strike AG. In a similar manner the dip of the second plane is laid off at EDF, except that in this case FE is made equal to CB to insure that both new strikes are the same distance below the given strikes. The new strikes CH and FH are prolonged until they meet at H, and GH is the plan or map view of the line of intersection. The bearing of GH may be measured with a protractor. HK is laid off at right angles to GH and made equal to FE and CB. Angle HGK is the true slope of the line of intersection.

3.54 Practice problems

See Chapter 8, Group 30.

3.55 Cut-and-fill problems

To establish the boundaries of cut-and-fill areas in highway construction or other operations involving earthwork, it is simply necessary to superimpose the proposed contours on the existing contours, mark the points where the two contour systems are in

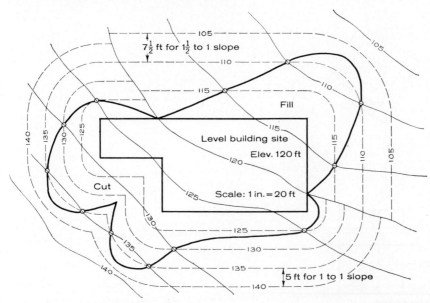

FIG. 3.37. Cut-and-fill boundaries.

agreement, and connect the marked points to outline the extent of the cut and fill.

Refer to Fig. 3.37. On the line where the level building site at 120 ft is crossed by the 120-ft contour there will be no cut or fill at all. The intersections of the 120-ft contour with the edges of the building site will mark the points of change from cut to fill. The proposed contours, shown by dashed lines, are laid off on the higher sides of the area 5 ft apart to give a 1 horizontal to 1 vertical slope for the cut. On the lower sides of the area the proposed contours are laid off $7\frac{1}{2}$ ft apart to give a $1\frac{1}{2}$ horizontal to 1 vertical slope for the fill. Where the proposed contours (dashed lines) cross the existing contours (solid lines) of equal elevation, boundary points for the cut and fill are established. These points are then connected by a smooth curve. The depth of the cut or the height of the fill at any point is the difference between the proposed and existing elevations as interpolated from the contours. This information can be used to calculate the cubic yards of earthwork.

3.56   Shades and shadows

Architects' drawings usually show the shades and shadows on a building or on the ground because they emphasize the third dimension and make the building look more natural. The principle of a line piercing a plane, or other surface, is all that is needed for determining shades and shadows, except careful visualization. The following brief explanation is given:

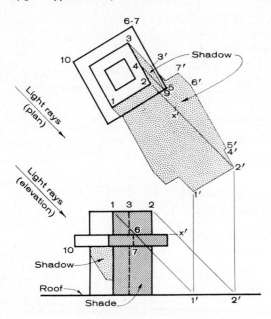

FIG. 3.38. Shades and shadows.

The source of light is assumed to be an infinite distance away. This makes the rays of light parallel in any view, and they may be assumed to come from any direction. Any portion of the object itself which the light rays do not strike is *in shade*. The *shadow* of an object cast on another surface is that portion of the surface which the light rays cannot strike because the object interferes. Or the shadow could be cast on the object itself by another part of the object. The *shade line* on an object is the boundary line between the light and dark, or shade, surfaces. The *shadow line* is the boundary line of the shadow.

The method of determining the shadow cast by a chimney onto a flat roof, or onto itself, and the shade on the chimney is shown in Fig. 3.38. The shade surface had a crosshatched boundary, and the shadows are spattered. The direction of the light rays in each view is assumed as shown. A light ray through point 1 in both views is seen to pierce the ground at 1′. One from point 2 hits the ground at 2′. The rays going over the top of the chimney from 2 to 4 hit the ground from 2′ to 4′. But the ray from point 4 is the last one that can hit the ground. Rays from 4 to 3 hit the projecting horizontal ledge and cast their shadow on the ledge. The shadow line 4′ to 5′ is cast by the upper edge of the ledge from 9 to 5. The ledge from 5 to 6 casts the line 5′ to 6′. The vertical corner 6 to 7 casts the shadow line 6′ to 7′. Thus the entire shadow in the plan is determined. The shadow in the front elevation is cast by the lower edge of the ledge going both directions from point 10.

It is extremely necessary to visualize the rays passing over the top of an object and around the sides in order to determine at all times what line on the object is casting the shadow. This can be done most easily by holding a pencil over the object like a ray of light and by moving it around to determine the casting line. Note that the ray from point 2 appears to pierce the horizontal ledge surface at x', which point is outside the chimney. Therefore that ray misses the ledge and goes on to hit the roof at 2′. But the ray from point 3 does hit the ledge at 3′. This must be watched to see which plane a ray strikes first.

The following simple rules will prove useful for checking the direction of any shadow line:

1. A line which is parallel to a plane will cast a shadow on that plane parallel to itself.

2. The shadow of a straight line cast onto any plane will pass through the point where the line itself pierces the plane.

It should be noted that in architectural work the light rays are usually assumed to follow the direction of the diagonal of a cube from the front upper left corner to the rear lower right corner of the cube; in other words, over the left shoulder of the observer. A different direction has been used in Fig. 3.38 so that the difference between shade and shadow can be illustrated clearly.

For shadows cast onto any other surface than a plane the method is the same except that the point where a line pierces a curved surface must be found by the methods explained in Chapter 6.

4

REVOLUTION

4.1 Change-of-position method

Two entirely different methods are in use for solving drafting-room problems. All the problems that have been explained in the foregoing chapters of this text have been solved by the direct or change-of-position method. When using this method the draftsman imagines the object to be in a fixed position; he never thinks of it as being moved or turned around. If he wishes to obtain a different view, he must imagine himself occupying a different position in space in order to see what he wishes to see on the object. The object which is being drawn always remains stationary while the observer moves around it. This method gives an easier and a more direct solution for most practical problems, and it is used, even though unconsciously, by the large majority of draftsmen.

4.2 Revolution method

The alternate method for solving drafting-room problems is the revolution method, which requires the draftsman to remain in a fixed position and to imagine the object turned around so that he may obtain any desired view of it. Although this is not the most practical method for drafting-room use, some problems may be solved more easily by this method. For this reason the student should become thoroughly familiar with both methods, and therefore should exercise his own judgment in selecting the one which is better suited to the solution of each particular problem.

Best results can be obtained by using a combination of the two methods, in order to avoid double revolution, a common source of error. The explanations which follow in this chapter are all based on this combination method.

4.3   Principles of revolution

Revolution is based on the following fundamental principles which must be thoroughly understood before any attempt is made to solve problems. The use of these principles should insure a clear conception of what actually happens in space when revolution is performed.

I. *A point, when revolving in space, always revolves about some straight line as an axis.*

Caution: Do not try to revolve any point until you see clearly just how the axis lies about which the point is to be revolved.

II. *A point always revolves in a plane which is perpendicular to the axis, and its path in this plane is always a circle.* The radius of this circle is the shortest distance from the point to the axis.

III. *This circular path of the point will always appear as a circle in the view in which the axis appears as a point.*

In the front elevation view of Fig. 4.1, the axis AB appears as a point (a_2b_2), and the path of the point X in revolving about the axis AB appears as a circle. The true length of the radius is always seen in this view.

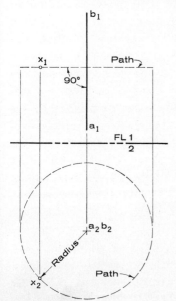

IV. *This circular path will always appear as a straight line at right angles to the axis in the view which shows the true length of the axis.*

Since the plane of this circular path is perpendicular to the axis, it must show as an edge in looking at right angles to the axis to see its true length. The plane view of Fig. 4.1 illustrates this principle. The circular path of the point X is shown appearing as a straight line at right

FIG. 4.1. Illustrating the principles of revolution.

angles to the axis AB, which is in its true length in this view. In this case the plane of the path of motion is vertical, being at right angles to a level line.

4.4 Revolution principles illustrated

The four principles that have been stated are fundamental, and they must be strictly adhered to whenever revolution is performed. When problems are to be solved in which the axis does not appear as a point in either of the views given, a new view must be drawn which will show the axis as a point. The change-of-position method of drawing is employed in order to obtain the two views which will show the axis in its true length and as a point. These two views will then show the axis in such simple positions that the revolution may be easily performed in accordance with the four fundamental principles.

Figure 4.2 illustrates the statements made in the preceding paragraph. The point X and the line AB are given in the plan and the front elevation and the point X is to be revolved about the axis AB. The axis appears in its true length (a_2b_2) in the front elevation, but it does not appear as a point in either of the views given. It is therefore necessary to draw a new view (view 3) which will show the axis as a point (a_3b_3). The revolution may now be performed by means of view 2 and view 3, in which the problem becomes exactly the same as the one shown in Fig. 4.1.

In some problems the axis does not even appear in its true

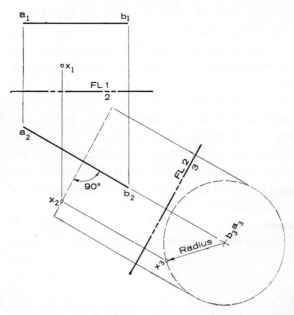

FIG. 4.2. A point revolving about a frontal line.

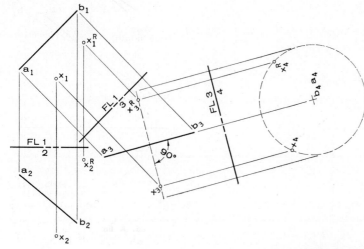

FIG. 4.3. A point revolving about any oblique line.

length in either of the given views. This is the case with the axis *AB*, which is given in the plan and front elevations of Fig. 4.3. The point *X* is also given, and is to be revolved about *AB* as an axis. It is then necessary to draw a new view (view 3) showing the axis in its true length (a_3b_3), and another view (view 4) showing the axis as a point (a_4b_4). The point *X* is also located in these two new views. This procedure again reduces the problem to its simplest form, that shown in Fig. 4.1, by performing the revolutions in views 3 and 4 only. If it is desired to stop the point *X* at some definite position as $x_4{}^R$ in view 4, the new position in all views is easily determined by projection lines and measured distances from folding lines, as was explained in previous chapters.

4.5 Practice problems

See Chapter 8, Group 31.

4.6 To find the true length of any line

Revolve the line about an axis which intersects the line and which is parallel to the image plane that will show the true length of the line. If it is desired to revolve the line *AB* in Fig. 4.4 until its true length will be shown in the front elevation, the axis must be taken parallel to the front image plane, and usually it is taken in its simplest position, which is vertical. The revolution is then performed exactly as in Fig. 4.1, the four fundamental principles being adhered to. In this case the vertical axis is taken touching one end of the line *AB*, and the other end of the line is revolved to the position $b_1{}^R$. The line *AB* now lies parallel to the front image plane; hence it appears in its true length in the front view.

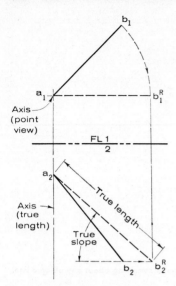

FIG. 4.4. A line revolved about a vertical axis.

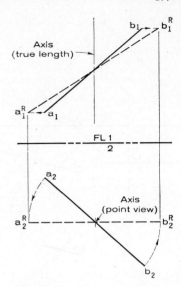

FIG. 4.5. A line revolved about a level axis.

The same process is followed if it is desired to show the true length of the line in any other view. In Fig. 4.5 the line *AB* is revolved so that its true length appears in the plan view, and the axis of revolution must be taken as a level line. This figure also illustrates the fact that the axis may intersect the line either at one end or at any point along the line; in the latter case both ends of the line are revolved. Strictly speaking, every point on the line is revolved that is not on the axis of revolution.

The true slope of a line will be seen in the elevation view that shows the true length of the line, provided the line has been revolved about a vertical axis, as in Fig. 4.4. *Revolution of the line about any axis except a vertical axis will change its slope.*

4.7   Practice problems

See Chapter 8, Group 32.

4.8   To find the true size of any plane

The plane must always be revolved about an axis that lies on the plane itself. This axis must also lie parallel to the image plane for the view which will show the true size of the plane. For example, if the plane is to be revolved so that it will show in its true size in the plan view, the axis must be taken as a line in the plane and parallel to the plan image plane; in this case, level. After this axis has been located, the procedure is that of previous problems, each point on the plane being revolved about the axis until the plane itself is level. In Fig. 4.6, the plane *ABC* is to be revolved to

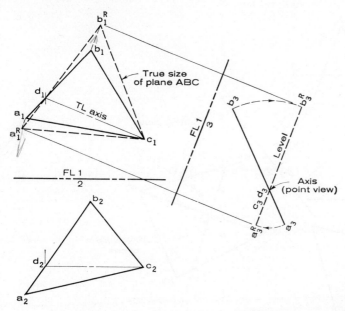

FIG. 4.6. True size of plane by revolving until level.

a level position. A level axis CD is chosen which lies in the plane. An extra view, 3, is drawn which shows the axis as a point (c_3d_3). By means of views 1 and 3, the solution is again reduced to the basic principles, as in Fig. 4.1. In view 3 the plane is shown revolved to a level position ($a_3{}^Rb_3{}^R$); while in this position, it shows in its true size ($a_1{}^Rb_1{}^Rc_1$) in the plan view.

In this same manner, any plane figure in any possible position in space may be revolved so as to show in its true size and shape in any desired view. Attention is called to the fact that a plane automatically appears as an edge in the view which shows the axis as a point, since the axis is on the plane. This edge view must always be seen before the revolution can be performed.

4.9   Practice problems

See Chapter 8, Group 33.

4.10   To find the true size of any dihedral angle

The dihedral angle between two planes is measured by a plane angle which is cut out of the two planes by a third plane passed perpendicular to the line of intersection of the two given planes. The line of intersection of the two given planes is first determined. Then a plane is passed at right angles to this line of intersection, and the intersection of this third plane with each of the given planes is found. These two lines of intersection form the plane angle which measures the dihedral angle. The plane of this angle is revolved to such a position that its true size may be seen.

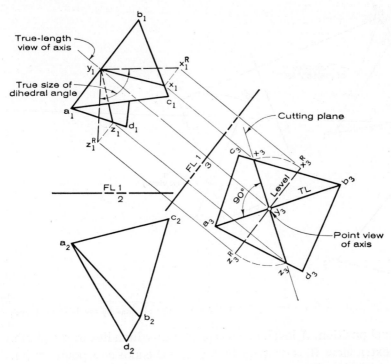

True-length view of axis

True size of dihedral angle

Cutting plane

Point view of axis

FIG. 4.7. Dihedral angle by revolution.

In Fig. 4.7 the plan and elevation views of planes ABC and ABD are given. It is desired to find the true size of the dihedral angle between these two planes. From inspection it may be seen that the line AB is common to both planes and therefore is their line of intersection. If the line of intersection were not given, it would have to be found by any of the conventional methods previously described. A new view 3 is drawn showing line AB in its true length. A cutting plane is drawn perpendicular to line AB at *any point,* such as Y. Since line AB is shown true length in view 3, the cutting plane appears as an edge in this view and cuts line YZ out of plane ABD and line YX out of plane ABC forming the angle XYZ. In order to see the true size of angle XYZ, the cutting plane is revolved to a level position using *any* level line in the cutting plane as an axis. In this case a level axis of revolution is passed through point Y, although it could be passed conveniently through either X or Z if desired. In the process of revolution, point Y on the axis remains stationary and X and Z move to new positions in accordance with the principles of revolution explained in Section 4.3 until all are in the same level plane. The true dihedral angle is shown in the plan view at the revolved position of angle XYZ.

4.11   Practice problems

See Chapter 8, Group 34.

4.12   To find the true size of the angle a line makes with a plane

The axis for performing the revolution for this problem *must be perpendicular to the plane.* If the line is revolved about any other axis, the angle the line makes with the plane will be destroyed. A view of the line and plane is drawn which shows the plane true size. The axis will appear as a point in this view. The line is revolved about the axis until it appears true length in the view which shows the plane as an edge, and the required angle is measured in this view.

In Fig. 4.8 the plan and elevation views of line XY and plane ABC are given and it is desired to find the angle between the given line and plane. Additional views are drawn showing plane ABC first as an edge and then true size. Notice that in view 4 it is not necessary to draw plane ABC, but it is necessary to show line XY. An axis perpendicular to plane ABC appears as a point in view 4. The axis chosen is passed through X, although it could be passed through Y or any other point on line XY if desired. Line XY is revolved until it is parallel to plane 3. The required angle is shown in view 3 between the edge view of ABC and the true length of XY. It should be noted that the distance from the

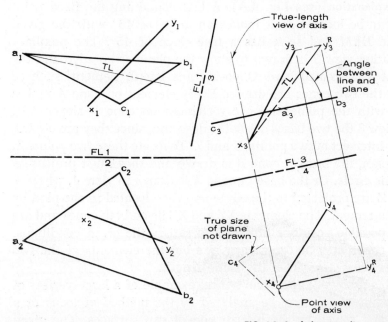

FIG. 4.8. Angle between line and plane by revolution.

point Y to the plane did not change during the revolution. This condition must always hold if the angle with the plane is to remain the same in space.

4.13 Practice problems

See Chapter 8, Group 35.

4.14 Cone locus

Analysis. Problems involving the intersection of two right cones of revolution arise when the position of a line in space is to be determined so as to satisfy two given angular conditions. Such problems are analyzed by thinking of the locus of the line for each condition. For example, if a line passing through some fixed point is to have a definite slope, it must lie on a cone of revolution whose vertex is the fixed point and whose elements all have the given slope. In other words, that cone is the locus of all lines, regardless of their length, that have the given slope and contain the given point. If the required line is to make a fixed angle with a given line or plane, another cone is drawn to show the locus of all possible positions of the line to satisfy this condition. Both cones *must have the same vertex and the same length elements* because they are both generated by the same line. The line of intersection of these two cones lies on both cones and therefore satisfies both given conditions.

Explanation (see Fig. 4.9). A line containing the fixed point X is to be located so it makes an angle of 45° with the given plane HKM and so it has a true slope of 45°. The auxiliary elevation view 3 is drawn to show the plane as an edge and the point X is located here. Choosing any length element, such as x_3r_3, the two cones are drawn. Every element on cone A makes 45° with the plane and every element on cone B slopes 45°. In view 3 the two bases are both edges and, since they are circles, they intersect at two points, Y and Z. To locate these two points in the plan, the base of cone B is drawn and they must project on to this circle. Or the base of cone A is drawn in view 4, where Y and Z are projected to the circle and then located in the plan by measurement. The elements XY and XZ lie on both cones and are two possible positions of a line to satisfy the given conditions. The cones may be drawn any size but the elements of the two cones must always be the same length.

This problem is just one typical example of a large variety of problems which may be analyzed by the method of locus of a line and solved by the intersection of two surfaces. The given

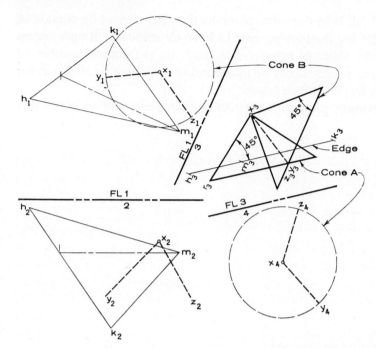

FIG. 4.9. Cone locus. Two right cones of revolution.

conditions fix the kind of surfaces the line must lie on so their line of intersection will satisfy these conditions. It is easily possible to have angular specifications given which would be impossible for any line in space to fulfill. In such a case the impossible situation is evident as soon as the two cones are drawn because they lie wholly outside each other and will never intersect.

4.15 Practice problems

See Chapter 8, Group 36.

4.16 Restricted revolution

Situations sometimes arise in practice where a study must be made to determine whether a rotating part on a machine is free to rotate through a complete circle or whether it will interfere with other parts of the machine. This study will either determine the actual amount of clearance or it will show how the design must be changed to give the desired clearance.

By the principles of revolution (Section 4.3) the farthest-out point on the moving part moves in a circle in a plane at right angles to the axis. The intersection of this circular plane with any possible interfering surface is determined. This line of intersection, which is in the same plane as the circle, must lie outside of the circle for the moving part to have clearance. If it cuts across the circle, there is interference, and the maximum number of degrees of rotation may be measured or data may be obtained for redesigning for full clearance.

Revolution principles furnish the simplest way to make this study.

5

VECTOR GEOMETRY

5.1   Introduction

A thorough understanding of vectors is important for all engineers because vectors are vitally involved in many engineering sciences such as mechanics, kinematics, and alternating-electric-current theory. The algebraic manipulation of vectors is dealt with in the branch of mathematics known as *vector analysis*. This chapter will describe the graphic manipulation of vectors, usually a faster method than the algebraic and sufficiently accurate for most purposes. Vectors can be added, subtracted, multiplied, and even differentiated either graphically or algebraically, but the treatment here will be limited to graphic addition and its practical applications.

A vector quantity is one which has both *magnitude* and *direction*. Force, velocity, and acceleration are examples of vector quantities because their complete description requires a statement of direction as well as size. For example, if an automobile is said to be traveling at 40 mph it is a statement of speed; but if it is said to be traveling 40 mph in a due north direction it is a statement of velocity, a vector quantity.

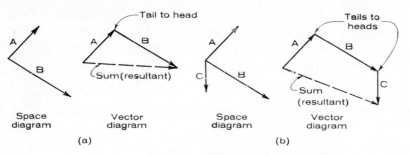

FIG. 5.1. Addition of concurrent coplanar vectors: (a) addition of two
coplanar vectors, (b) addition of three (or more) coplanar vectors.

A vector is represented by a line whose length is proportional
to the magnitude of the quantity and whose direction is the same
as the direction in which the quantity is acting. An arrowhead is
placed at the end of the vector to complete the identification of
the direction.

5.2   Addition of concurrent coplanar vectors

When two or more vectors lie in the same plane they are said to be
coplanar. If they act through the same point they are *concurrent,* literally "running together." The graphic addition of such
vectors is a simple matter of placing the vectors together *tails to
heads,* maintaining their given magnitudes and directions as illustrated in Fig. 5.1, and connecting the tail of the first vector
with the head of the last vector to find the sum of the vectors.
When several vectors are added the sequence of addition will not
affect the result, although it is desirable to proceed in some pre-
conceived order to avoid confusion. If the direction of the sum,
or closing vector, is from the tail of the first vector to the head of
the last vector it is called the *resultant,* and the resultant will
have the same effect as the total effect of the individual vectors
that have been added. If the direction of the resultant is reversed it is called the *equilibrant,* and the equilibrant will have
the effect of opposing the added individual vectors to place the
system exactly in equilibrium.

Figure 5.2 shows the addition of vectors by the parallelogram
method. A careful examination of this figure will reveal that the
same result is obtained as in the "tails to heads" method just described. Since the parallelogram method is a
mechanical one which has no advantages in

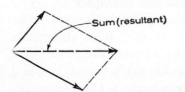

FIG. 5.2. Addition by parallelogram method. Not recommended.

logic or clarity and which, indeed, may lead to confusion when there are more than two vectors to be added, the student is advised to learn thoroughly and rely exclusively on the "tails to heads" method for the work in this chapter.

5.3   Practice problems

See Chapter 8, Group 37.

5.4   Addition of concurrent noncoplanar vectors

When vectors do not all lie in the same plane they are *noncoplanar*. The theory of addition is the same as for coplanar vectors, except that three dimensions are involved instead of two and the principles of multiview or orthographic projection must be used to describe and solve the problem which now occupies three-dimensional space.

   Figure 5.3 illustrates the addition of noncoplanar vectors. Notice that a plan and elevation view of the given vector quantities, a *space diagram*, is drawn, and a plan and elevation view of the addition process, a *vector diagram*, is drawn. Each vector in the plan view of the vector diagram must be parallel to the corresponding given vector quantity in the plan view of the space diagram, and each vector in the elevation view of the vector diagram must be parallel to the corresponding given vector quantity in the elevation view of the space diagram. The scaled length of a vector must be laid off in a view which shows the vector true length. If a vector is oblique (not parallel to either principal

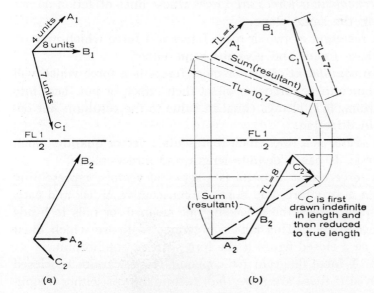

(a)   (b)

FIG. 5.3. Addition of concurrent noncoplanar vectors: (a) space diagram and (b) vector diagram.

plane of projection), its scaled length must be laid off in a true-length view obtained either by the auxiliary view method or by revolution.

5.5   Practice problems

See Chapter 8, Group 38.

5.6   Forces in structures

The most practical application of vectors occurs in solving forces acting in engineering structures. Quite a different problem is presented here. In Sections 5.2 and 5.4, it is usually required to find only the value and direction of one unknown, namely, the resultant or the equilibrant. In structures there may be one or more known forces (or loads) but the values of two coplanar or three noncoplanar unknowns, of given direction, must be determined. This requires the use of special methods which will now be illustrated. The principles already learned in Chapter 3 are all that are needed for these solutions.

5.7   Definitions

The following is a summary of definitions already given but repeated here in words which apply especially to forces.

1. *Concurrent forces* are forces whose lines of action all pass through a common point.

2. *Coplanar forces* are forces whose lines of action all lie in the same plane.

3. *Noncoplanar forces* are forces whose lines of action do not all lie in the same plane.

4. A *resultant* of two or more forces is a force which may replace these forces and give the same effect.

5. An *equilibrant* of two or more forces is a force which will just balance these forces, or offset their effect, or put them into equilibrium. It is always equal in value to the resultant but opposite in direction.

6. A *vector* is a line which represents a vector quantity, such as a force. It has a definite length and direction.

7. A *force vector diagram* is a series of vectors, representing forces in a structure, laid down in consecutive order, and each one starting where the preceding one stopped, or tails to heads as in Section 5.2. This makes a connected figure which must always be a closed figure if the forces are in equilibrium.

Note: A force diagram for coplanar forces cannot be closed and solved if there are *more than two* unknowns, either magnitudes or directions.

5.8   Solution of coplanar forces

Figure 5.4a shows a space drawing, to scale, of a boom carrying a load of 2,000 lb on the cable B and held in position by the cable A. The pin D, through which three forces are acting in the same plane, is held in equilibrium by these three. Figure 5.4b is an equilibrium, or free-body, diagram of the pin D with the three forces shown acting on it which hold it in equilibrium. This diagram is not drawn to scale and is used only to determine whether the unknown forces are tension or compression forces. Tension forces always act away from the joint and compression forces toward the joint in this diagram.

The force in the cable B is known, but the forces in the cable A and in the boom C are unknown and are to be determined. These three forces may be represented by vectors, which, when laid down one after another in any order, will form the closed vector diagram shown in Fig. 5.4c.

In order to facilitate the drawing of the vector diagram, a system of lettering known as Bow's notation is used. In this system each space in the space drawing between two external forces (or loads) or between an external force and the frame is given a capital letter. Each space inside the frame or between two members is given a lower-case letter. An arrow is then drawn to indicate which way it is desired to read these forces around the pin D. It is immaterial which way the arrow reads, but, when it is once

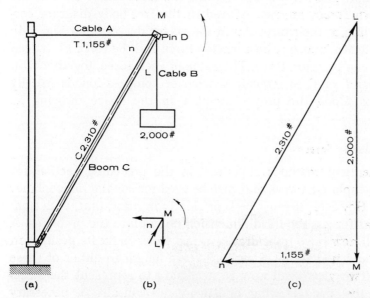

FIG. 5.4. Graphical solution of coplanar forces: (a) space drawing, (b) free-body diagram, and (c) vector diagram.

chosen, all forces should read around the pin in the same direction as the arrow. With the arrow pointing as shown in Fig. 5.4a, the vertical force reads LM, the level force reads Mn, and the slanting force reads nL. If the arrow had pointed the other way the sequence of the notation would be reversed. The vertical force would read ML, the slanting force Ln, and the level force nM.

The vector diagram may now be drawn, starting with the vertical force, because it is the only force whose value and direction are both known. This force is drawn as a vector in the vector diagram parallel to its line of action in the space drawing, and is measured equal to 2,000 lb to some scale. This force acts down and, since its name is LM, it is labeled L at its beginning and M at its end. The next vector in order, following the direction of the arrow, is Mn, which is drawn from M parallel to the line of action of the level force in the space drawing. The length of this vector Mn is unknown. However, the next vector in order is nL, which has a fixed direction and must return to L to close the diagram. Therefore the vector nL can be drawn through L, and point n is thereby determined. The vector diagram is now completed, and, since the entire diagram lies in one plane and shows in its true size, the two unknown vectors may be scaled to determine their values. Arrows are placed along each vector in the vector diagram, all arrows reading around the diagram in the same direction, which direction is determined by the arrow on the known vector LM. The arrows representing the direction of each of the unknown forces are next placed in the free-body diagram corresponding to their direction in the vector diagram. In Fig. 5.4b the force Mn is seen to be a tension force and the force nL is seen to be a compression force. The value of each force, together with a capital T or C to indicate tension or compression, is usually recorded along the proper member in the space drawing as shown.

5.9 Noncoplanar forces

The identical methods explained in the preceding section for solving coplanar forces may also be used for solving noncoplanar forces. However, it must be kept clearly in mind that, with noncoplanar forces, the third dimension enters into the problems. A line will now have to be drawn in two views to fix its position in space, and it may not show in its true length in either of these views. Two views will now be necessary to represent the structure in the space drawing, and two views will also be necessary to completely represent the vector diagram. It is therefore most essential that the reader grasp the idea that there are two separate and distinct objects side by side in space, namely, the structure itself and its corresponding vector diagram, which is a closed figure having each line parallel in space to the respective member of the structure in which it acts. Figure 5.5 shows a photograph of a small model of a hanging frame and its corresponding vector diagram. Section 5.11 gives a complete explanation of the solution for this same frame.

Briefly stated, the solution consists of drawing a plan view and some elevation view of the structure itself and its vector diagram, as they stand side by side, and finding the true length of each line of the closed vector diagram. Before a detailed solution is presented, several statements will be given for the express purpose of checking up the ability of the reader to think clearly and properly in three dimensions, and of directing this ability toward the solution of three-dimensional vector diagrams.

5.10   Principles

The following basic principles regarding the relations existing between orthographic views and vector diagrams are stated below without proof. They should be carefully studied and thoroughly understood before the solution of a problem is attempted.

1. The vector diagram for any system of forces in equilibrium must be a closed figure.

2. A three-dimensional vector diagram will appear as a polygon in all orthographic views.

3. Each line, or vector, of a vector diagram in space lies parallel to the corresponding member in space.

4. If two lines are parallel in space they will appear parallel in all orthographic views.

5. In the same view of any space drawing and its corresponding vector diagram, each line of the vector diagram will be parallel to its corresponding member in the space drawing.

6. The true force in any member is shown in the vector diagram only when the member itself appears in its true length in the space drawing.

7. The true length of each line of the vector diagram must be seen before it can be scaled.

8. All vertical vectors show in their true lengths in all elevation views.

9. All horizontal vectors show in their true lengths in the plan view.

10. The true lengths of all other vectors may be found by orthographic methods.

11. Noncoplanar problems usually have one or more known forces and three forces whose directions are known but whose magnitudes are unknown. If more than three items are unknown, either directions or magnitudes, the solution is impossible. If a structure contains more supporting members than are absolutely necessary for stability, it is statically indeterminate and must be solved by other methods than those described here.

12. With three unknown forces, the easiest solution is made possible when two of these unknown forces appear in the same line or when one unknown force appears as a point in some view. The effect of this procedure is to make one view of the vector diagram a triangle instead of a four-sided figure, and it is this one principle that is most fundamental to the solutions presented here.

13. The two unknown forces which appear in the same line in some view *must be kept consecutive in the vector diagram,* and should be the last two forces to be drawn in the vector diagram.

14. If one unknown force appears as a point, it may occur anywhere in the sequence of forces in the vector diagram.

15. Bow's notation should be used *in one view only, never in two views.*

5.11   Solution of special case

Figure 5.6 shows the plan and front elevation views of a simple hanging frame carrying a 1,000-lb load suspended at the point *A.* This frame is shown in the photograph in Fig. 5.5. It is desired to

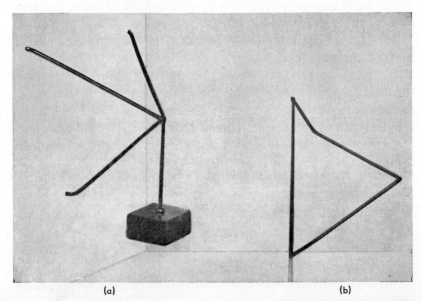

(a)   (b)

FIG. 5.5. (a) Structure with a load. (b) Vector diagram.

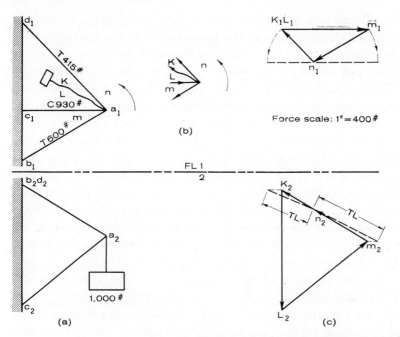

Force scale: 1" = 400 #

(b)

FIG. 5.6. Graphical solution of noncoplanar forces: (a) space drawing,
(b) free-body diagram, and (c) vector diagram. Compare with Fig. 5.5.

solve the three unknown forces acting in the members *AB*, *AC*,
and *AD*. This is a special case, because two of the members, *AB*
and *AD*, appear in the same line in one of the given views.

The plan and front elevation views of the structure itself are
first drawn to a definite scale, as in Fig. 5.6a. This is the space
drawing. The spaces between the members are lettered using
Bow's notation, which is placed on the plan view only. The vertical load appears as a point in this view and it is shown bent to
one side just for convenience in placing notations around it. The
true direction of the line of action of the force in this member is
not altered by this temporary bending. In this case, if it were bent
to the right it would lie between the two forces in *AB* and *AD*
which must be kept consecutive and the solution would be much
harder. Therefore the vertical load is bent as shown. (See Section 5.10, principle 13.) The curved arrow is placed in the plan
view to give the order of reading the forces around the joint *A*.

The vector diagram is then drawn to some assumed scale,
starting with the only known vector *KL* and drawing it in both
the plan and elevation views, as in Fig. 5.6c. Since the vector *KL*
is vertical, it will appear as a point in the plan and in its true
length in the front elevation. Each vector should be lettered at
both ends and in both views as soon as it is drawn. The next vector in order, *Lm*, is drawn from *L* in both views of the vector dia-

gram. Its direction in each view of the vector diagram must be parallel to the direction of the member AC in the corresponding view of the space drawing. But the length of the vector Lm is unknown and therefore the position of m is still undetermined.

The remaining two vectors in order are mn and nK. These two forces appear in the same line in the front elevation of the space drawing, and their vectors must therefore have the same direction in the front elevation of the vector diagram. Since they are consecutive, they will actually be on the same line in the front elevation of the vector diagram and may be drawn in that view as one line, to close the diagram back to the starting point K. This determines point m in both views but still leaves point n undetermined. Since the point m is now known in the plan view, the remaining two vectors, mn and nK, may be drawn in this view to close the diagram back to the starting point K. This can be done because the direction of both vectors in the plan view is known. The intersection of these two vectors determines point n, which may also be projected to the front elevation and located there. The four-sided vector diagram is now closed and complete in both views, and it is therefore a definitely fixed figure in space as shown in the photograph in Fig. 5.5. It remains only to find the true length of each line or vector in the diagram and to scale that length to determine the value of each force. Arrows should be placed to indicate the direction of action of each force in each view of the vector diagram, as in Fig. 5.6c.

A free-body diagram, showing the forces acting on the joint at A, should be drawn as shown in Fig. 5.6b. This sketch is just a reproduction of a portion of the view of the space drawing in which the notation has been placed. The plan view was used in this case. The actual member is shown broken off and the line of action of the force acting in that member is also shown. Arrows, to indicate the direction of action, are placed on each force in the free-body diagram exactly as they point in the corresponding view of the vector diagram. The forces nK and mn, acting away from the member in the free-body diagram, are therefore tension forces and the force Lm is a compression force. The value of each force, with T or C to indicate tension or compression, should be placed along the member in which it acts in one view of the space drawing. The plan view was chosen for recording these values in this problem, because it would show them more clearly.

Note: If Bow's notation is used in this way, any force which is designated by one or more lower-case letters acts in a member of the structure, and any force which is designated by two capital letters is an external force, or a load.

5.12   Solution of general case

Attention is called to the fact that in the problem which has just been solved in Section 5.11 the frame was placed in such a position that two of its members appeared in the same line in the front elevation. This may be considered to be a special case, for not every structure might happen to have two of its members appearing as a line in either of the given views. However, by the method of Section 2.12 for finding the edge view of any plane, it is very easy to draw a new elevation view of any two intersecting lines, in which they will appear as the same line.

In order to solve a general case, a new elevation view of the structure must first be drawn in which two of the members appear in the same line. The plan and this new elevation view are treated exactly the same as though they were the given views, and two views of the vector diagram are drawn corresponding to these two views. From this point the solution is exactly the same as for the special case in Section 5.11, and for that reason the drawing for this solution in detail is not given.

Any possible general case having three unknowns may be reduced to the special case and solved as such by simply drawing some new view showing two of the unknowns in the same line. If there should happen to be more than one known load, as in airplane work, these known forces should always be kept consecutive and should be drawn first in the vector diagram. In some cases it might be easier to use two elevation views or possibly an elevation and an inclined view instead of the plan and elevation views. Any two related orthographic views will enable the problem to be solved by this method, provided the vector diagram is represented in two views corresponding to those chosen for the space drawing.

5.13   Solution by seeing one unknown as a point

The plan and front elevation of a frame are given as in Fig. 5.7 (p. 100). It has a horizontal load and the member AB is level, showing in its true length in the plan. A new elevation, view 3, is drawn of the entire frame so as to show the member AB as a point. Bow's notation is placed in the plan. Two views of the vector diagram are constructed to correspond with views 1 and 3 of the space drawing. The known force MN is drawn first and then the force Nh, which appears as a point in the elevation. The two closing forces can then be drawn in the elevation, fixing the point K, which is projected to the plan. The plan can then be completed and the vector diagram closed. The problem is completed from here just the same as the one in Fig. 5.6.

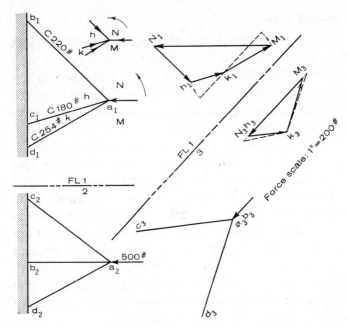

FIG. 5.7. Solution by seeing one unknown as a point.

The solution could also have been made with a view showing AC (or AD) as a point, but an extra view showing its true length would be necessary before it could be seen as a point. Generally speaking, this method should be used if one of the members shows in its true length in one of the given views; otherwise, the method of Section 5.12 should be used.

5.14   **Practice problems**

See Chapter 8, Group 39.

5.15   **Solution by finding the true size of the plane of any two unknowns**

This method involves finding the edge view and true-size view of a plane containing any two unknowns and locating the other unknown and the load line in both these views. The length of the load vector would be given. A summation of forces is then taken perpendicular to the plane. The only two forces which have a component perpendicular to the plane are the load and the one unknown which is not on the plane. Therefore, for equilibrium, their two components which are perpendicular to the plane must be equal in value and opposite in sense. Since the load component is known, the other component is known also, and therefore the stress in that member is known. The other two unknowns may be solved by completing the vector drawing in the true-size view of the plane.

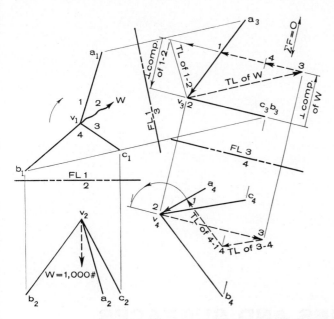

FIG. 5.8. Solution by true size of plane of two unknowns.

In Fig. 5.8 is shown a tripod structure carrying a vertical load of 1,000 lb suspended at V. View 3 shows the edge view of the plane containing the members to B and C, and view 4 shows the true size of this plane. The member to A and also the load line are both located in these two views. The vector drawing is now started in views 3 and 4, using Bow's notation. The 1,000-lb load is laid off from 2 to 3 in view 3 since this vector is in its true length here. Otherwise the true length would have to be found and the load laid off on the true length. This determines the length of the load vector 2 to 3 and fixes the length of the component of this force which is perpendicular to the plane. The component of the force 1 to 2 is equal in value and opposite in sense to the component of the load just determined. This establishes point 1, and the vector drawing 1-2-3-4 may now be completed in views 3 and 4. The true length of 1-2 is found by revolution. Both 3-4 and 4-1 already show in their true lengths in view 4.

In this method the vector drawing is superimposed on the views of the structure. This makes a fast and easy solution and does not sacrifice clearness. In either view it is easily seen that all three forces act toward the joint V, and therefore the members are all in compression.

5.16   Practice problems

See Chapter 8, Group 39 (by the method in Section 5.15).

6

CURVED LINES AND SURFACES

6.1 Introduction

The majority of the objects that an engineering draftsman is called upon to draw are bounded by straight-line corners and plane surfaces. All of the problems which might arise in working with these can be handled by the methods that have been explained in the previous chapters.

However, many situations confronting a draftsman require a knowledge of curved lines and curved surfaces for the correct representation of an object or for the proper solution of a problem. Many varieties of cams, gears, and screw conveyors depend entirely on curved lines for their very operation. Modern concrete construction often includes a warped surface on a dam, an irrigation canal, or a syphon. Many varieties of hydraulic and irrigation pipe problems include combination surfaces made up of planes, cones, and cylinders.

An engineer should be able to tell, at a glance, whether or not it is possible to develop a certain surface. He should also know how to develop any surface which can be developed. In order to have a thorough foundation in the fundamental subject of drawing, he must at least be familiar with curved and warped surfaces and with the solution of problems relating to them. The material in this chapter is not presented with any intention of covering this subject completely. Sufficient explanations are furnished to give a reading knowledge of the various types of surfaces and to enable one to solve the ordinary problems with which he might be confronted. The major emphasis will be placed upon problems dealing with cylinders and cones, because these occur more frequently in engineering practice.

In general, the change-of-position method is still the most natural one to use. However, in some cases, such as finding the true length of a series of lines, the revolution method will be found to be very much easier and faster. Judgment should be exercised as to which method is the more efficient for each solution attempted.

6.2   Curved lines

Curved lines fall naturally into two general classes, which are determined entirely by the third dimension. These two classes are *lines of single curvature* and *lines of double curvature.*

6.3   Single-curved lines

Lines of single curvature lie wholly in one plane and are therefore just two-dimensional or plane curves. There may be an infinite number of these curves, determined by an infinite number of mathematical equations. Some of the most common are the circle, ellipse, parabola, and hyperbola. These four curves are called conic sections. Then there are the trigonometric curves, the most common of which are those of the sine, cosine, and tangent. Involute and cycloidal gear teeth and irregularly shaped cams furnish other practical illustrations of single-curved lines.

The detailed explanations for the actual construction work involved in drawing single-curved lines are omitted here, because they are usually taught in elementary drawing courses and because they may be found in almost every current text on engineering drawing.

6.4   Double-curved lines

Lines of double curvature do not lie wholly in the same plane. They are three-dimensional or space curves. There may also be an infinite number of these curves, which are determined by an infinite number of mathematical equations. Most of these curves do not have a practical physical application but occur more often in pure mathematical calculations. The only curve of this class which will be discussed further will be the helix.

6.5   Helix

The helix is a vital part of practically every machine in existence. The statement may be safely made that there is no automobile, airplane, ship, train, engine, or power-driven machine that does not have a helix somewhere in its structure, in the form of either a screw thread or a coil spring.

The helix is generated by a point moving around an axis and parallel to that axis at the same time. The point usually (though not necessarily) moves at a uniform rate in both directions. If it remains a fixed distance away from the axis, a *cylindrical helix* is generated, since the point will be traveling around a cylinder. If the point moves on the surface of a cone so it goes around the axis as it approaches the vertex of the cone, it generates a *conical helix*. A conical spring is an example.

Figure 6.1 shows the plan and elevation of a right-hand cylindrical helix which is generated by a point starting from A and traveling once around the cylinder whose diameter is D while it is traveling a distance L parallel to the axis of the cylinder. The distance L is called the *lead* of the helix. The *pitch* of a helix on a double, triple, or quadruple screw thread is the distance from a point on one thread to the corresponding point on the adjacent thread.

The curve is drawn by dividing the distance around the cylinder into any number of equal parts and by dividing the lead into the same number of equal parts (16 in this case). As the

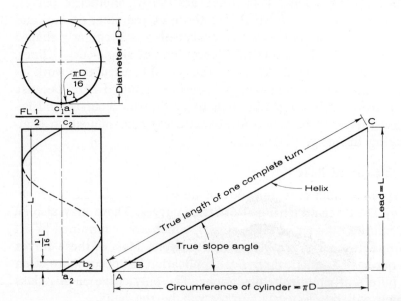

FIG. 6.1. Right-hand cylindrical helix.

point *A* travels one-sixteenth of the distance around the cylinder it travels one-sixteenth of the lead, gaining the position at *B*. By taking each position separately the construction is easily made, and the entire curve may be drawn as shown.

If a triangular paper is cut out, like the one shown in Fig. 6.1, having a base equal to the circumference of the cylinder and an altitude equal to the lead of the helix, the hypotenuse of the triangle is the actual or true length of one complete turn of the helix. The angle shown is the true slope angle of the helix, and the tangent of this angle is seen to be $L/\pi D$. When the triangular paper is wrapped around the cylinder the hypotenuse coincides with the helix, because the helix has a uniform slope and must in consequence be a straight line when it is developed. This triangle gives a clearer idea of the helix than can be obtained by any other means.

6.6   Practice problems

See Chapter 8, Group 40.

6.7   Curved surfaces: definitions

1. A *surface* is the path of a moving line (except when the line is straight and moves endwise like an arrow).

2. The *generatrix* is the line which moves to generate a surface.

3. An *element* is the generatrix at any one position. A curved surface may therefore be considered as being made up of an infinite number of elements, or different positions of the generatrix. (An element is also considered to be any straight or curved line on the surface.)

4. A *directrix* is a line which guides or defines the motion of the generatrix. There may be one or two directrices for the same surface.

5. A *director* is a plane to which the generatrix remains parallel.

6. A *ruled surface* is one which may be generated by a straight line. A straight-edged ruler may be laid on a ruled surface so that it will touch the surface for its entire length.

7. A *double-ruled surface* is a surface through any point of which it is possible to have two intersecting straight-line elements.

8. A *single-curved surface* is a ruled surface that can be developed or rolled out into a plane.

9. A *warped surface* is a ruled surface that cannot be developed.

10. A *double-curved surface* is a surface which can be generated only by a curved line and which has no straight-line elements.

11. A *surface of revolution* is the path of any line which revolves about a straight line as an axis. The line that revolves may be straight or curved. A straight line revolving in this manner will generate a single curved surface of revolution if it is parallel

Table 5. Outline of curved surfaces

		Name	Generatrix		Developable or not
			Kind of line	Kind of motion	
Ruled:	Single-curved	Cylinder	Straight line	Touches single-curved line; remains parallel to straight-line directrix	Yes
		Cylinder of revolution	Straight line	Revolves about straight-line directrix to which it is parallel	Yes
		Cone	Straight line	Touches single-curved line; intersects straight-line directrix	Yes
		Cone of revolution	Straight line	Revolves about straight-line directrix which it intersects	Yes
		Convolute	Straight line	Remains tangent to any double-curved line	Yes
		Helical convolute	Straight line	Remains tangent to a helix	Yes
	Warped	Helicoid	Straight line	Touches helix and its axis; makes constant angle with the axis	No
		Hyperbolic paraboloid	Straight line	Touches two nonintersecting nonparallel straight lines; remains parallel to a plane	No
		Conoid	Straight line	Touches one straight line and one curved line; remains parallel to a plane	No
		Cylindroid	Straight line	Touches two curved lines; remains parallel to a plane	No
		Hyperboloid of revolution of one sheet	Straight line	Revolves about an axis which is nonintersecting and nonparallel	No
Double-curved:	Surfaces of revolution	Sphere	Circle	Revolves about its diameter	No
		Torus or annulus	Circle	Revolves about any straight line except its diameter	No
		Ellipsoid (prolate)	Ellipse	Revolves about its major axis	No
		Ellipsoid (oblate)	Ellipse	Revolves about its minor axis	No
		Paraboloid	Parabola	Revolves about a symmetrical axis through its focus	No
		Hyperboloid of two sheets	Hyperbola	Revolves about an axis through both foci	No
	Irregular	Unnamed	Any curved line	Moves along any other curved line	No

to or intersects the axis, giving a cylinder of revolution or a cone of revolution respectively. But if the straight line is neither parallel to nor intersects the axis, the surface generated will be a warped surface called the hyperboloid of revolution of one sheet (or one piece).

12. A *right section* of a surface of revolution is a plane section which is perpendicular to the axis; it must always be a circle.

Table 5 gives a classification of most of the surfaces commonly occurring in engineering work, and it will show quickly the general class to which a surface belongs, the way in which it is generated, and whether or not it can be developed.

6.8   Cylinder

A cylinder is generated by a straight line moving around another straight line and always remaining parallel to it. It is usually considered to be a closed surface, which means that the generating line returns to the place from which it started. The path of any point on this generating line may be any curved line and does not have to be a circle. If this path should be a circle in a plane perpendicular to the axis of the cylinder, the surface would be a *cylinder of revolution*, or just a round cylinder as in Fig. 6.2.

Most cylinders in practice are round, as in the case of tanks, pipes, etc. For this reason the common conception of cylinders

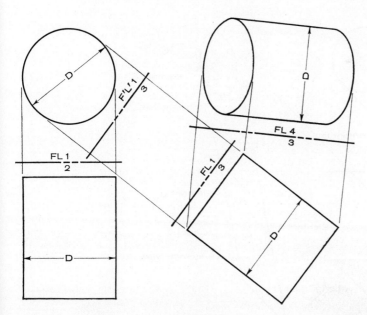

FIG. 6.2. Cylinder of revolution.

is that they have to be cylinders of revolution. This is a false conception. The cylinder is a cylinder of revolution only when the right section is a circle. If an oblique section of a cylinder is a circle, the shape of the right section of the cylinder is an ellipse, as in Fig. 6.3. This cylinder is an elliptical cylinder.

6.9   Representation of a cylinder

Although a cylinder is considered to have an infinite number of elements, only two elements are drawn in any view and these are the two that project in the extreme outside position on each side. They are called the extreme elements. The cylinder is represented in any view by drawing its upper or lower base, or both bases, and the two extreme elements in that view. This fact is shown in Fig. 6.3. The extreme element A in the front view is not an extreme element in any other view. If two views of a cylinder arc related and one of those views is the true-length view, then the extreme elements in either view will be over the centerline in the other view. In Fig. 6.3 notice the location of elements B and C, which are located by projection and folding-line measurement.

In order definitely to locate any element on a cylinder for purposes of projection, some point on the element is always used which lies in one of the given bases or in any other plane section of the cylinder. This one point may be projected to different

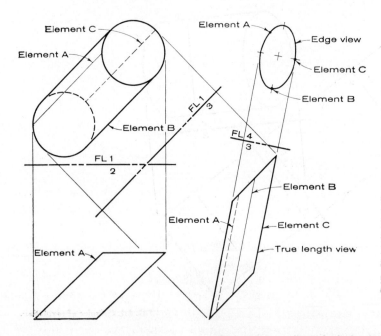

FIG. 6.3. Oblique elliptical cylinder showing extreme elements.

views, and the element may then be drawn. It is necessary to locate only one point on each element in any view, because all the elements project parallel to the axis in every view.

Since all the elements on the surface of a cylinder are parallel to the axis, the following theorem may be stated.

Theorem 10. The view of a cylinder which is drawn showing the axis as a point will show the cylinder surface as an edge and will give the true shape of its right section (see view 1, Fig. 6.2, and view 4, Fig. 6.3).

Also it may be proved that any oblique view of a cylinder of revolution will show it always to have the same width. This fact is a very useful one and is restated as a theorem.

Theorem 11. A cylinder of revolution will have a width equal to its diameter in every possible view (see Fig. 6.2).

6.10   Practice problems

See Chapter 8, Group 41.

6.11   To find where a line pierces a cylinder

First method: by auxiliary views

Analysis.   A view of the cylinder is drawn looking parallel to its axis and showing the cylinder as an edge (by Theorem 10). The given line is then located in this same view. The two piercing points where the line passes through the cylinder will be apparent in this edge view and may be projected back to all other views. The solution by this method is not shown, but it would be complete in view 4 of Fig. 6.3 if the given line was located in that view.

Second method: using only the two given views

Analysis.   It is usually easier to solve problems in curved surfaces by using straight-line elements on the surface. They should be used whenever it is possible. In order to cut straight-line elements on the cylinder, a cutting plane will have to be passed parallel to the axis and to all the straight-line elements. This cutting plane will have to contain the given line also, and it will contain two straight-line elements where it cuts the surface of the cylinder. These two elements lie in this same plane with the given line, which will have to either intersect them or be parallel to them. If it happens to be parallel to them it is parallel to the cylinder and cannot pierce it. If it intersects them, the two points of

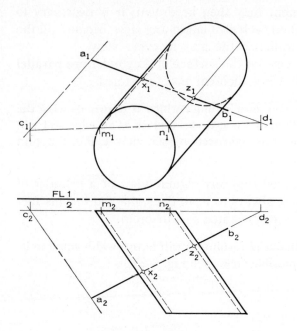

FIG. 6.4. Line piercing a cylinder.

intersection are the required piercing points. The given views
will always allow one to determine by inspection whether or not
the line is parallel to the cylinder (by Theorem 1, Section 3.2).

Explanation (see Fig. 6.4). The oblique cylinder and the line
AB are given in the two views, and the points at which the line
AB pierces the cylinder are required.

In order to use straight-line elements on the cylinder, a plane
is passed containing the line AB and parallel to the axis of the
cylinder. The line AC is the auxiliary line used, and the cutting
plane BAC is the desired plane. This cutting plane intersects the
plane of the upper base of the cylinder in the line CD and cuts
across this base of the cylinder at the points M and N. These two
points are the upper ends of the two straight-line elements, which
also lie in the plane BAC. The two elements are drawn in both
views, since they have to be parallel to the axis; the points where
they intersect the line AB are the required piercing points X and
Z. These two points, which have been determined independently
in both views, should also check by projection between views.
The plane of the lower base could also have been used for determining the two straight-line elements.

6.12   Practice problems

See Chapter 8, Group 42.

6.13   Plane section of cylinder with vertical axis

First method: by auxiliary views

Analysis.   A new elevation view is drawn showing the given plane as an edge and showing the cylinder as well. This new view will indicate the points at which each element of the cylinder pierces the plane or is cut off by the plane. This piercing point on each element can then be located on the proper element in any other view by projection. When all the piercing points have been located in any view, the entire curve may be drawn.

Explanation (see Fig. 6.5). The cylinder and the plane *ABCD* are given in plan and front elevation. The new elevation is drawn looking parallel to the level line *AB*, in order to show the plane as an edge. The cylinder is also shown in this view. The cylinder is divided into 16 elements equally spaced around the circumference, the division being made in the edge view of the cylinder. It is usually safer to number these elements (in pencil) while the problem is being solved. These 16 elements are located in the new view, and it is apparent where each element is cut off by the plane. These piercing points are all located in the front elevation. A line connecting them gives the plane section across the cylinder; in this view it is a curved line, as shown. Any point upon it will be visible if it lies on a visible element of the cylinder. Exactly half of the elements will be visible in any view; they will be the ones that are closest to the observer.

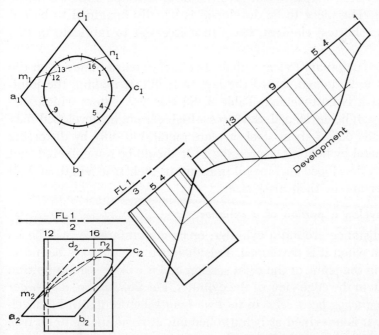

FIG. 6.5. Plane section and development of a cylinder.

Second method: using two views only

Analysis.   By the method of finding where a line pierces an oblique plane the point may be determined at which each element on the cylinder pierces the plane. A line connecting these points will give the desired curve that the plane cuts on the cylinder.

By the method of Section 3.21, vertical projection planes are drawn containing one or two elements at a time. One of these planes is shown at *MN*, taken so as to contain the two elements numbered 12 and 16. The points at which these two elements pierced the plane are found in the front view and marked with two small circles. The piercing points of all 16 elements are found in the same way; they determine the required curve.

Both of the methods just explained are applicable to all cylinders, whether they are cylinders of revolution or not.

6.14   Practice problems

See Chapter 8, Group 43.

6.15   Development of surfaces

To develop a surface means to lay it out into a plane like a flat piece of metal. All containers that are to be made out of sheet or plate metal must first be laid out on the metal when it is in flat sheets. The curves where the metal is to be cut must all be laid out on the surface so that, after the plate has been cut and rolled or bent into shape, it will give the desired shape and size to the vessel. The piece to be developed is usually imagined to be cut on the shortest element, because it is easier to fabricate in this way.

It is customary to leave about two inches extra material on the edges to be joined, called the lap, to facilitate making the joint, unless it is butt-welded. Table 5, Section 6.7, shows which surfaces can be developed. Surfaces which cannot be developed may be made out of sheet steel or plate metal, but only by distorting the metal by heating or pressing it. It should be remembered that on any developed surface all lines are in their true length and all angles are in their true size.

6.16   To develop a portion of a cylinder

The distance around a cylinder, or its circumference, will be its length when it is developed. This distance must always be measured in the plane of the right section and it will be seen in its true length in the edge view of the cylinder. For this reason, *a cylinder must always been seen in its true length* before it can be developed. It is imagined as being rolled out at right angles to its true length. The elements will always remain parallel to each other, and the distance between them is taken from the view in which they appear as points. With all the elements located, the end of each element may be projected from the true-length view to give the curve desired.

Explanation (see Fig. 6.5). It is desired to develop the portion of the cylinder lying above the plane. Since both elevation views show this cylinder in its true length, the development may be made from either one. In this case it is made from the auxiliary elevation, because the cut on each element is more easily obtained in this view.

In the development the 16 elements are spaced their true distance apart as seen in the plan, and are drawn parallel to their true length. The end of each element is projected over from its true length to the proper element, as is shown on element 13. The correct developed curve is drawn by connecting the ends of the elements with a fair curved line.

When developing a surface, one must be able to place the metal outside or inside up as required. If outside up, the metal would have to be bent or rolled down. Developments are made either way, depending on the kind of machine that is to do the bending. The following is a sure and easy method of placing the outside of the metal up: In the true-length view of the cylinder in Fig. 6.5, select the numbered element that is closest to you, which is No. 5. Then element No. 4 is on the right of No. 5, and it must also be on the right in the development. For placing the metal inside up, No. 4 would be placed on the left of No. 5. When developing larger pieces, the use of a larger number of elements is recommended for more accurate results.

6.17   **Practice problems**

See Chapter 8, Group 44.

6.18   **To represent an oblique cylinder of revolution cut by a level plane**

Analysis.   The position of the axis of the cylinder will have to be fixed. It is apparent that the cut will show as an ellipse in the plan view. A new elevation view showing the true length of the axis is drawn first, and the extreme elements of the cylinder may be drawn in all views (by Theorem 11). In this new elevation the level plane will appear as an edge, and the cut made across the cylinder by this plane will give the major axis for the elliptical cut in the plan view. The minor axis will be the full diameter of the cylinder. The ellipse may be drawn by using the two axes thus obtained.

Explanation (see Fig. 6.6). The axis AB of the cylinder is given in the plan and front elevation views. The level plane is also fixed and contains the point B. The new view 3 shows the axis AB in its true length, and the cylinder is drawn in all three views. The cut of the level plane across the cylinder in view 3 determines the major axis for the ellipse, and the minor axis is already known. The ellipse is drawn in the plan by use of the two axes and the trammel method (see Appendix A.2).

This elliptical cut may also be obtained by drawing the edge view of the cylinder, view 4, selecting elements such as X and Y, and finding where each element is cut by the plane as in Section 6.13. The element method is more basic, but the axis method is much faster. After the ellipse has been drawn, it should be checked by projection between plan and front elevation.

If this cylinder is to be developed, it should be rolled out from view 3 and the elements spaced equally in view 4.

6.19 Practice problems

See Chapter 8, Group 45.

6.20 To represent an oblique cylinder of revolution cut by a frontal plane

Analysis. The axis of the cylinder must be given. It is apparent that the cut will show as an ellipse in the front elevation. The solution will be exactly the same as for the preceding problem,

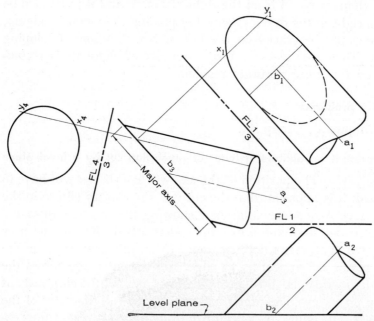

FIG. 6.6. Cylinder cut by a level plane.

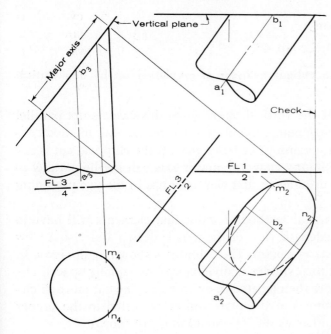

FIG. 6.7. Cylinder cut by a frontal plane.

except that the true-length view of the axis will be taken from the front elevation. This is the case because the given cutting plane is parallel to the front image plane and therefore shows in its true size in the front elevation and as an edge in the true-length view of the cylinder.

Explanation (see Fig. 6.7). The axis *AB* of the cylinder and the vertical plane are given. View 3 shows the true length of the axis *AB* and the edge view of the vertical or frontal plane passing through the point *B*. The cylinder is then drawn in all views, and the cut across the cylinder by the plane in view 3 determines the major axis, as in Section 6.18. The ellipse is drawn by the trammel method. It must project between views as before.

View 4 is also shown and is used for solving the problem by the element method. The solution for the elements *M* and *N* is shown. The development would be conceived as rolling out the cylinder from view 3.

The most practical method of developing a cylinder of this nature is to square up the paper so that the true length of the cylinder, as in view 3, is perpendicular to the T square. Then the developed surface rolls out parallel to the T square and the elements are all at right angles with T square. By this method the development may be placed in a square or symmetrical position on a separate sheet of paper.

6.21 Practice problems

See Chapter 8, Group 46.

6.22 To represent an oblique cylinder of revolution cut by any vertical plane

Analysis. If the vertical plane, as in this case, is not parallel to the front image plane, the major- and minor-axes method cannot be used for obtaining the front view. If the elliptical cut is not in its true size in any view, the major axis will not lie parallel to the axis of the cylinder in that view as it has in the two preceding problems.

The fundamental method of individual elements will have to be used. A true-length view of the axis is drawn and also a view showing the axis as a point. The cylinder is shown in all views. In the edge view, the cylinder is divided up into equally spaced elements, which are then located in every view. The cut on each element is determined in the plan and is projected to the proper element in all other views. The solution is not shown.

6.23 Practice problems

See Chapter 8, Group 47.

6.24 To represent any oblique cylinder cut by any oblique plane

Analysis. The cylinder and one of its bases will have to be given in both views, as well as the oblique plane. In general, the easiest method of solution is to obtain a new view of both showing the plane as an edge. In this view it will be apparent where each element on the cylinder pierces the plane. The elements may be projected from one view to another by using the given base of the cylinder. However, this process does not place the cylinder in a position to be developed. A view showing the true length of the cylinder, with both ends of each element, would still have to be drawn, and the cylinder would have to be rolled out from this true-length view.

If the true-length view were to be drawn first, the method of Section 3.21 would have to be used to find where each element of the cylinder pierced the plane. This method would place the cylinder in a better position for developing, but it would cause more work in finding the ends of the elements.

The solution is not shown.

6.25 Practice problems

See Chapter 8, Group 48.

6.26   To draw a plane tangent to a cylinder from a point not on the surface

Analysis.   A plane that is tangent to a cylinder will contain one element of the cylinder and will therefore be parallel to all the other elements on the cylinder. This plane will also intersect the plane of the base of the cylinder in a line which is tangent to the base of the cylinder. A line through the given point and parallel to the cylinder will lie in the required plane. From the point where this line pierces the plane of the base of the cylinder a line may be drawn tangent to the base (on either side). This line and the line containing the given point are two intersecting lines which determine the required plane. There are two possible solutions, giving two different planes.

A good check method for this problem is to draw an edge view of the tangent plane after it has been determined. In this view the plane as an edge must coincide with an extreme element of the cylinder in the same view. The solution is not shown.

6.27   Practice problems

See Chapter 8, Group 49.

6.28   The cone

A cone is generated by a straight-line generatrix moving around a straight-line directrix which it intersects. The angle between these two lines may vary.

A cone of revolution is a special type generated when the generatrix revolves about the axis, thus making a constant angle with the axis. The cone of revolution occurs more commonly in engineering practice. It is found in gearing, in roller bearings, and in friction drives in machinery. Cones other than cones of revolution are found in pipe reducers, offsets, and other metal pieces. Since, by definition, the generatrix intersects the axis, all the straight-line elements on a cone pass through the vertex. Every cutting plane which cuts straight-line elements on the cone must contain the vertex of the cone.

A cone is different from a cylinder. It is impossible to draw any view of a cone which will show it as an edge. It is also impossible to draw any one view that will show the true length of all the elements, as can be done with the cylinder. As a rule, the easiest way to solve all cone problems is to use cutting planes giving straight-line elements. As in the cylinder, only one point on the base needs to be determined to locate straight-line elements. The other end of all straight-line elements is the vertex of the cone.

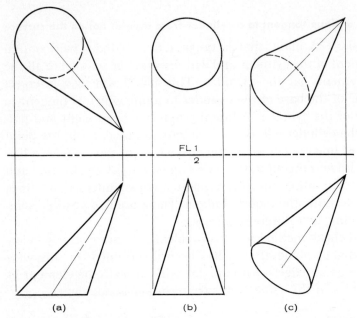

(a)   (b)   (c)

FIG. 6.8. Representation of cones.

6.29   Representation of a cone

A cone is usually represented by showing its vertex, two extreme elements, and some plane section which is called the base. If the vertex is omitted, two plane section cuts must be shown as well as two extreme elements. The cone in Fig. 6.8a is elliptical-shaped. On this cone exactly half the elements are visible in the front view, and more than half are visible in the plan.

When the axis of a cone appears as a point in some view, only the base of the cone is shown in that view, as in the cone of revolution in Fig. 6.8b. In this cone all the elements are visible in the plan view, and exactly half the elements are visible in the front view.

Figure 6.8c shows a right cone of revolution with the axis at an angle with both image planes. The base is perpendicular to the axis and is a circle. In the plan view more than half the elements are visible, and in the front view less than half.

6.30   Practice problems

See Chapter 8, Group 50.

6.31   To find where a line pierces a cone

Analysis.   If the given line pierces the cone, it must intersect two straight-line elements on the surface. Both these elements contain the vertex of the cone. A plane may be drawn so as to

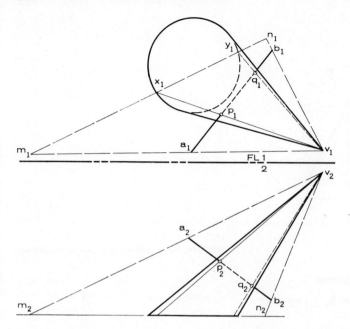

FIG. 6.9. Line piercing a cone.

contain both the given line and the vertex of the cone. If this
plane cuts through the cone at all, it will cut across the base and
will cut two straight-line elements on the surface of the cone.
The base end of these two elements is determined by the intersection of this plane with the plane of the base of the cone. The
two elements may then be drawn to the vertex in both views and,
since they are in the same plane with the given line, they will intersect that line and give the required piercing points.

Explanation (see Fig. 6.9). The line *AB* and the oblique cone
are given in two views. The plane *ABV* is drawn so as to contain
the line *AB* and the vertex *V* of the cone. The plane produced is
found to intersect the plane of the base of the cone in the line
MN, and it actually cuts across the base of the cone at the points
X and *Y*, which are the lower end of the two required elements.
These elements are drawn to the vertex and are found to intersect the line *AB* at the points *P* and *Q*. These two points are on
the line *AB* and on the cone surface, and they are therefore the
required piercing points. They may be determined independently
in both views and checked by projection.

The solution as given above is completed in the two given
views. After the plane *ABV* is drawn, a new view may be drawn
showing that plane as an edge and passing through the vertex of
the cone. This plane cuts across the base of the cone in the same
line of intersection, *MN*, as was obtained by the use of only two
views. From this point the procedure will be the same.

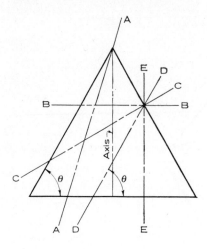

FIG. 6.10. Plane sections of a cone of revolution.

6.32   Practice problems

See Chapter 8, Group 51.

6.33   Plane sections of a cone of revolution

Explanation.   The cone of revolution of Fig. 6.10 is shown cut across by five different cutting planes.

Plane *A*, which passes through the vertex, cuts an isosceles triangle. The cuts made by the four other planes are known in mathematics as the conic sections, which were referred to in Section 6.3.

Plane *B*, being at right angles to the axis of the cone, cuts out a circle.

Plane *C*, which is not perpendicular to the axis and which cuts all the elements, cuts out a true ellipse.

Plane *D*, which is parallel to one element and which has the same slope as this element, cuts out a parabola.

Plane *E*, which makes any angle greater than θ with the level plane, cuts out a hyperbola.

The true size of each cut may be seen in the view showing the true size of the plane in which it lies. Individual points may be determined by drawing a number of straight-line elements and finding where each element is cut by each plane. It is also possible to determine individual points by using a series of right-section cutting planes which would cut circles from the cone and straight lines from the given planes.

6.34   Practice problems

See Chapter 8, Group 52.

6.35   To develop a cone of revolution

Analysis.   A cone of revolution is usually developed by using the vertex and some base that is a right section. Since every point on the right-section base is equidistant from the vertex, the locus of all these points in the development will be a portion of a circle with its center at the vertex and with a radius equal to the slant height of the cone. The length of this circle along the arc will be the same as the distance around the base circle. A series of equally spaced elements may be drawn on the cone and also on the development. If the elements happen to be cut off before they reach the vertex, the true length of each element must be found from the vertex (or the base) to the cut end. These lengths, when laid off on the proper elements in the development, establish the points for the desired curve.

Explanation (see Fig. 6.11). A portion of a cone of revolution is given with a right-section base and a sloping top. Since the vertex is missing, the cone is extended so that the vertex can be used. In the plan view the lower base circle is divided into 16 equal spaces, giving equally spaced elements which are drawn to the vertex and numbered in both views. The slant height R is the radius for the developed base circle, and the length of the circular arc is made equal to the sum of the 16 equal spaces around the base circle in the plan. The 16 elements are then

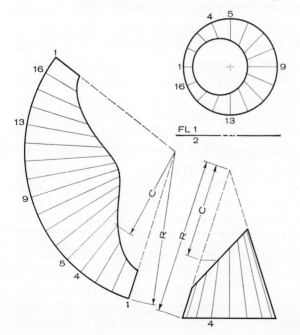

FIG. 6.11. Development of a cone of revolution.

drawn in the development. To obtain the curve for the top cut, the true length of each element from the vertex to the top cut is found, in this case by revolution. The figure shows this process for one element (4), whose true length, C, is found by revolution. All the elements of this cone, when revolved, will coincide with the extreme element. Accordingly, it is necessary only to project the upper end of each element to an extreme element to obtain its true length. This true length is then laid off from the vertex along element 4 in the development, to give the point on the upper curve. The ends of all the elements found in this manner determine the top curve. The cone should be cut on the shortest element as shown.

If it had been desired to develop only the portion of the cone above the sloping plane, it would still have been much easier to develop the entire cone to the lower base or to some other right section. This is the easiest way to space the elements in the development, and when they are properly spaced the true lengths to any point on the element are easily determined and laid off.

The method which has just been explained will apply in developing any cone of revolution cut by any oblique plane.

6.36   Practice problems

See Chapter 8, Group 53.

6.37   To develop any cone when the vertex is available on the drawing

Analysis.   It is assumed that the given cone is not a cone of revolution. Then the elements all have different lengths, and the base, whether it is a right section or not, will not roll out as a circle. The surface will have to be divided up into elements and the triangulation method used. By this method, the true lengths of two adjacent elements and the true distance between their base are laid down in the development to form a triangle. This is repeated for the adjacent triangle, and so on for the entire surface. This method assumes that the surface between two adjacent elements is a plane surface, which is really not the case. However, with elements taken very close together, the surface becomes so nearly a plane that the error is negligible.

Explanation (see Fig. 6.12). The oblique cone is given as shown with a level, circular base. The vertex is available on the paper. The level base is divided into 16 equal spaces in order to give elements whose lower ends are spaced an equal distance, R, apart. The elements are all drawn to the vertex. The true length[1] of the shortest element, V-1, is found and is laid down to

[1] See Appendix A.3 for sheet metal workers' method of finding true lengths.

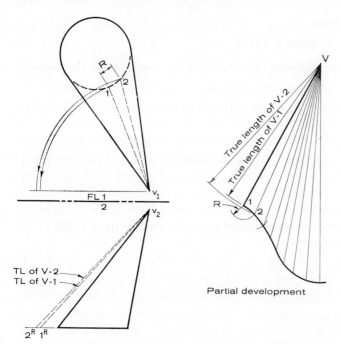

FIG. 6.12. Development of an oblique cone by triangulation.

start the development. The true length of the next adjacent element, V-2, is then struck off as an arc from V. The point 2 must lie somewhere in the development at a distance equal to the true length of V-2 from V. But the point 2 must also lie a true distance of R from the point 1. With the distance R for a radius, an arc is drawn with point 1 as a center. The point 2 must lie on both of the arcs just drawn, or at their intersection. The entire curve is obtained in the same manner. The figure shows the development of only half of the cone.

The development may be placed in any desired position on the sheet or even on a separate sheet. If it is desired to place the development square with the sheet, the longest element should be laid down first at right angles with the T square. Other elements will then be laid on both sides of the longest one.

If the given cone had been only a partial cone, it would have been produced to its vertex and treated in exactly the same way. The statement of this problem assumes that, if the cone is produced, the vertex will be available within the limits of the drawing paper. The triangulation should be made from the vertex to the base which is farthest from the vertex. The curve for the nearest base may be easily located after the elements are all placed in the development and after the curve for the farthest base has been completed.

6.38   Practice problems

See Chapter 8, Group 54.

6.39   To develop any cone when the vertex is not available on the drawing

Analysis.   Since the vertex is not available on the drawing, the cone will have to be given with an upper and a lower base. The conical surface is assumed to be composed of a series of trapezoidal areas which, if they could be produced to the vertex, would be triangles as in the previous problem. These areas lie between two adjacent elements and a section of each of the upper and lower bases. A diagonal is drawn across this trapezoidal area to divide it into two triangular areas. True sizes of these triangles are then found and laid down in consecutive order to obtain the development.

Explanation (see Fig. 6.13). The cone is given as shown, having both upper and lower bases level. The surface is first bisected by a vertical plane containing the axis of the cone. This plane locates the element BC and the two points B and C from which each base is divided into the same number of equal parts. The corresponding elements are then drawn between the two bases, as the element AD. These elements will all meet at the vertex in space. They give the first elemental trapezoidal area, $ABCD$, to be developed. The diagonal is drawn from A to C. The triangle ABC is then developed exactly as in Section 6.37. The triangle ACD is developed next, and the rest of the surface in

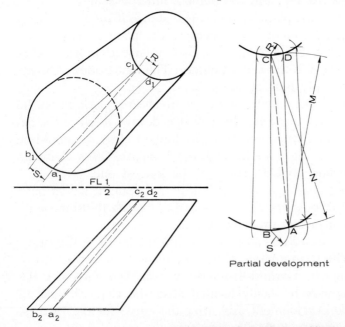

Partial development

FIG. 6.13. Development of an oblique cone without using the vertex.

the same way. The radii N and M are the true lengths of AC and AD, respectively. The radii R and S are the true lengths of the distances between equally spaced elements on the upper and lower bases, respectively. See Section A.3 in Appendix for special method to use here for finding true lengths.

Again the assumption has to be made that the trapezoid and the triangle are plane surfaces, which is not quite true. However, it affords an approximation sufficiently close for all practical purposes.

This problem would be more easily solved by the method of Section 6.37 if the surface could be produced to a vertex. In some cases it is impossible to get the vertex within the limits of the paper or the available steel plate, and then the method given in this section must be used.

6.40   Practice problems

See Chapter 8, Group 55.

6.41   To draw a plane which is tangent to a cone and contains a given point not on the surface of the cone

Analysis.   A plane tangent to a cone will contain one straight-line element on the cone and will contain the vertex. Since it must contain the given point, it will contain a line from that point to the vertex. The intersection of this tangent plane with the plane of the base will be a line tangent to the base of the cone. The point at which the line connecting the given point with the vertex pierces the plane of the base is one point on the line of intersection tangent to the base. From this point the line of intersection may be drawn tangent[1] to the base of the cone (on either side), giving two possible solutions. The required plane is represented on the drawing by its intersection with the plane of the base of the cone and the line from the given point to the vertex. If the line from the given point to the vertex should not pierce the plane of the base of the cone, it must be parallel to that plane. The tangent lines to the base would both be parallel to the line through the vertex. The solution is not shown here.

6.42   Practice problems

See Chapter 8, Group 56.

6.43   The convolute

The convolute is a single curved surface which is generated by a straight line moving so as to remain tangent to any double curved line.

---

[1] See Appendix, Section A.4, for drawing a line tangent to an ellipse.

The helical convolute is a convolute that is generated when the double curved line is a helix.

The generating line is sometimes assumed to extend (in all its different positions) to a given level plane and sometimes it is assumed to remain a constant length. Both these conditions are illustrated in Fig. 6.14.

The most practical application of the convolute is the helical convolute which is found in the blade of a screw conveyor. In this case the tangents to the helix do not extend to a level plane but only to a larger concentric cylinder inside of which the blade is to move.

Mathematicians call this surface a developable helicoid, a designation which is not strictly correct. The helicoid and the convolute are very similar, but Table 5 shows them to be in entirely different classes.

The convolute is a developable surface, but only within certain limits, which are determined by the slope of the helix. If this slope is exactly 45°, the convolute can be developed in one piece for only 1.41 turns around the cylinder. For angles of slope less than 45°, it can be developed from 1.0 to 1.41 turns about the cylinder in one piece. For angles of slope between 45 and 90°, the number of turns which can be developed varies from 1.41 to an infinite number. The number of turns it is possible to develop any helical convolute may be expressed in the form of an equation as follows:

$$N = \frac{1}{\cos A}$$

where N is the number of possible turns and A is the true-slope angle of the helix in degrees (see Section 6.5, Fig. 6.1).

6.44 Representation of a helical convolute

Explanation (see Fig. 6.14). The helical convolute is represented in a drawing by showing the helix and its cylinder, and several positions of the tangent to the helix. In Fig. 6.14 the given helix is drawn as in Section 6.5, by means of 16 equally spaced points. Figure 6.14 also shows the helix unwrapped from the cylinder in order to show its true length and true slope and to find the length of the tangents in the plan view. The solid curved lines show the convolute extending only to a larger concentric cylinder and for one complete turn around the cylinder. The dash lines show (only partially in each view) where the convolute would be if it extended to the level plane of the base of the cylinder.

To draw the convolute, at any point of the helix, such as C, a tangent is drawn to the helix. In the plan this is drawn tangent

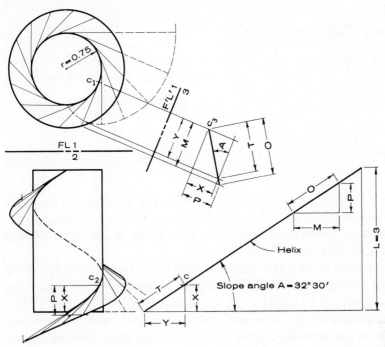

FIG. 6.14. Convolute.

to the cylinder. If it is to extend only to the larger cylinder, the plan-view length is immediately seen to be equal to M. If this distance is laid off level under the unwrapped helix, the difference in elevation between the ends of the tangent is seen to be equal to P. This fixes the front view of this tangent, and all other positions in this view are obtained by using this constant difference in elevation.

If it is desired to show the tangent extending to the level plane, it is apparent in the front view that the tangent would meet that plane at an elevation of X below point C. Projecting C over to the unwrapped helix fixes the plan view of the tangent as a distance Y in length, because the slope of the tangent is the same as the slope of the unwrapped helix.

This is shown a little clearer in view 3, which shows the tangent meeting the level plane when its length is T and going on through to the outer cylinder when its length is O. This view shows the tangent in its true slope and true length, the plan-view length, and gives the difference in elevation between the ends for both conditions which have been illustrated.

6.45   Development of the helical convolute of Fig. 6.14

Explanation (see Fig. 6.15). It is possible to develop a convolute by assuming it is composed of a series of triangles and by laying down the true size of these triangles in successive order

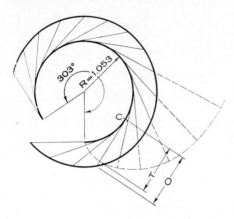

FIG. 6.15. Development of a convolute.

This is a cumbersome method, and there is another method which is very much simpler.

Since the helix has a constant curvature and a constant slope it will be a circle in the development of the convolute with a radius R, which is larger than the radius r of its cylinder. It has been proved by mathematics that the radius of this developed circle is equal to the radius of the helix cylinder divided by the square of the cosine of the slope angle of the helix.

$$R = \frac{r}{\cos^2 A}$$

Using this equation, the radius of the developed helix circle is calculated to be 1.053 and the circle is drawn. The distance the helix extends along this circular arc is equal to its slope length, or true length. This true length of the unwrapped helix is divided into 16 equal parts, and these parts are stepped off around the circle. At each of the division points on the circle a tangent is drawn, as at C, and is laid off in its true length (either T or O) from the helix. The ends of the tangents give the developed curve for either condition.

If the screw-conveyor surface as limited by the larger cylinder in Fig. 6.14 is to be developed, only one tangent needs to be drawn and laid off in its true length. The outside end of this tangent will determine the radius of the outside helix circle.

In practice the convolute will always be a surface limited by two concentric cylinders. The use of the formula greatly simplifies this development. The problem which has just been explained has assumed only one turn around the cylinder. The development shows that it is possible to develop a little more than one turn. The exact number of turns for the convolute shown, as calculated by the formula in Section 6.43, is found to be 1.186. It can also be easily calculated that one turn around the cylinder developed to 303° of the circle in the flat. This result is found by multiplying 360° by the cosine of the slope angle of the helix.

Attention is called to the fact that, theoretically, the convolute surface is the only one that can actually be developed without distortion for the purpose of making a screw-conveyor blade. However, this surface does not give a blade which is perpendicular to the shaft. Such a blade is to be preferred in practice, and the method for obtaining it will be explained in the next section.

6.46   Practice problems

See Chapter 8, Group 57.

6.47   The helicoid

The helicoid is a warped surface which is generated by a straight line moving about an axis in such a way that any two points on the line describe concentric helices having the same lead. The generating line need not intersect the axis, although it nearly always does. When it intersects the axis it also makes with it a constant angle. If this angle is equal to 90°, the surface is a right helicoid. Examples are the wearing surface of a square screw thread, a winding ramp (or chute) of constant grade, or a screw-conveyor blade. If the angle with the axis is an acute angle, the surface is an oblique helicoid, which would be found on the surface of a V-thread screw. Figure 6.16 shows a partial elevation of a cylindrical tower with a winding ramp and hand-rail going up around the outside. The ramp surface is a right helicoid.

The helicoid is represented by showing several positions of the generatrix and the helix with its axis. The length of the generatrix is limited, in all practical applications, by the given helix and

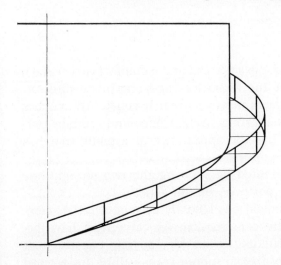

FIG. 6.16. Helicoid.

by another larger or smaller concentric helix which will be fixed by the conditions of the problem. These two limiting helices are shown in Fig. 6.16 together with a few positions of the generatrix. The screw-thread surfaces are so common that they are not shown.

Theoretically, it is impossible to develop a helicoid. Wherever this surface occurs in practice, made of steel plate or concrete, it must be made by distorting the metal or by approximating the surface by dividing it up into small sections. Most screw-conveyor blades are actually built by making an approximate development of a right helicoid, since the lead is usually comparatively small. The radius R of the developed inside helix may be calculated by the formula $R = r/\cos^2 A$ (see Section 6.45), and thus the developed helix circle may be drawn. The true length of this helix for one complete turn may be determined graphically (see Fig. 6.1) or by calculation, and this length laid out on the developed helix, which is the arc of a circle. The outer helix may be developed in the same way and the plate cut at both ends on radial lines, forming a segment of a circle. In shop practice it has been found that these radial lines will have to be trimmed or sheared off just a trifle to make a good fit. Also the plate will have to be hammered to make it come perpendicular to the axis, as a right helicoid should. This method of building a screw-conveyor blade is very similar to the convolute method of Section 6.45. It is usually followed in practice because of the simple method of laying out the blade and because of the desirability of having the blade surface square with the axis.

6.48   **Practice problems**

See Chapter 8, Group 58.

6.49   **Hyperbolic paraboloid**

The hyperbolic paraboloid is a warped and a double-ruled surface which has two straight-line directrices and one plane director. The straight-line generatrix moves so that it constantly touches two nonparallel, nonintersecting straight lines and remains parallel to some plane. It has the appearance of a plane that has been twisted.

The surface is represented by showing the two straight-line directrices and several positions of the generatrix.

The hyperbolic paraboloid has recently become a rather common surface in concrete construction. As carpenters have become more expert in building forms for concrete construction, engineers have felt more free to submit plans calling for warped

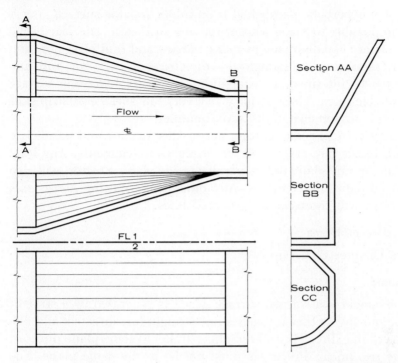

FIG. 6.17. Hyperbolic paraboloid.

surfaces if these will give either a better appearance to the structure or a higher operating efficiency. Consequently, we often find
this surface on concrete dams, irrigation ditches, tunnels, wing
walls, piers, etc.

In certain metal offset pieces this surface is approximated, or
really avoided entirely, by connecting the opposite ends of the
straight-line directrices with a diagonal straight line. This divides
into two triangular-shaped planes the surface which otherwise
would be a hyperbolic paraboloid.

In Fig. 6.17 is shown a very practical illustration of the oc-
currence of this surface in irrigation work. The open ditch of
section *AA* must be gradually changed into a syphon having a
closed section *CC* in order to conduct the water down into a deep
ravine and up to the proper elevation again on the far side of the
ravine. The section *AA* is first changed to the square section *BB*
and then to the section *CC*. The portion between the sections
AA and *BB* is shown in Fig. 6.17. It is called a transition, and its
sloping sides are hyperbolic paraboloids. Between sections *BB*
and *CC* there would be more surfaces of this same kind. The
plane director for the surface shown is a level plane, and the two
linear directrices are the vertical and the sloping lines at the ends
of the surface.

The hyperbolic paraboloid is an undevelopable surface, and it is impossible to make a form for this surface in one piece. The forms are usually made by using narrow and thin pieces of lumber fastened to the straight-line directrices at each end and supported by intermediate straight-line supports, since the surface is double-ruled. These strips give a very fair surface. A strip skiff is built in somewhat the same manner.

Any vertical plane cutting across this surface and parallel to both linear directrices will cut straight-line elements. Any level plane, or any plane that is parallel to the plane director, will also cut straight-line elements. Any other plane cutting across this surface will intersect it in curved lines.

6.50   Practice problems

See Chapter 8, Group 59.

6.51   Conoid

The conoid is a warped surface which is generated by a straight line moving so that it touches one straight-line directrix and one curved-line directrix and remains parallel to some plane director. The two linear directrices must not lie in the same plane. The curved-line directrix may be either an open or a closed curve.

This surface is represented by showing both linear directrices and at least the extreme positions of the generatrix, in all views. Sometimes several other positions of the generatrix are shown.

Figure 6.18 shows three views of a conoid which has a vertical

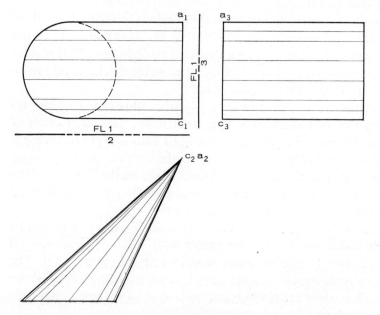

FIG. 6.18. Conoid.

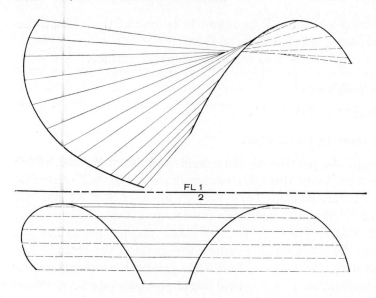

FL 1
2

FIG. 6.19. Cylindroid.

plane parallel to the front image plane for its plane director. Its curved directrix is a circle, in this case, and its straight-line directrix is the level line *AC*. Several random positions of the generatrix are shown in each view.

The conoid has very few uses in engineering practice. It is an undevelopable surface.

6.52   Practice problems

See Chapter 8, Group 60.

6.53   Cylindroid

The cylindroid is a warped surface which is generated by a straight line moving so that it touches two curved-line directrices and remains parallel to a plane director. The two curved-line directrices must not lie in the same plane. They may be either closed or open curves.

The cylindroid is represented by showing the two curved-line directrices and several positions of the generatrix. Figure 6.19 shows two views of a cylindroid having a level plane director. In order to show this surface more clearly the several positions of the generatrix are shown as solid elements on the upper face and as dashed elements on the under side of the surface. The twist or warp of this surface is clearly indicated in the plan view.

The cylindroid, like the conoid, has very few direct applications in engineering practice. However, these two surfaces, together with the hyperbolic paraboloid, occur in naval architecture on ship bodies and also on airplane fuselages. Nearly every streamlined surface will be found to be one of these three surfaces or some combination of them.

The cylindroid is also an undevelopable surface.

6.54   Practice problems

See Chapter 8, Group 61.

6.55   Special cases and limitations

The hyperbolic paraboloid, the conoid, and the cylindroid, which have just been explained, are so closely related to each other that it is sometimes difficult to classify them. Each surface has two linear directrices and one plane director to govern the motion of the generatrix. The definition of each surface prescribes the nature of the guides which control the travel of the straight line in generating that surface. And yet, a generatrix moving according to the definition for a conoid might also generate a hyperbolic paraboloid at the same time. To prove this last statement, a curved line may be assumed to lie somewhere on a hyperbolic paraboloid. If a generating line were to follow this curved line and one of the straight-line directrices, it would generate both a conoid and a hyperbolic paraboloid. In the same way it can be shown that a generatrix could move so as to satisfy the definition for a cylindroid yet could generate a conoid at the same time. These are possible cases, but they would occur only when the linear directrices had certain special shapes and certain relative positions.

It is easy to assume the two linear directrices in such positions that it is impossible for the generatrix to touch them both and remain parallel to the plane director. The two directrices and the plane director cannot be fixed arbitrarily. Even if only the linear directrices were fixed in space, it might be impossible to determine the plane director so as to generate the desired surface and traverse the entire length of both the linear directrices.

The extent of these three surfaces, and therefore their use, is very much limited. In practice, the desired surface is generated as far as possible, and a new surface of the same kind, or even a surface of an entirely different nature, may have to be used to complete the desired piece.

6.56   Hyperboloid of revolution of one sheet

The hyperboloid of revolution of one sheet (or of one piece) is a warped surface which is generated by revolving one straight line about another straight line as an axis, provided these two lines are not parallel or intersecting.

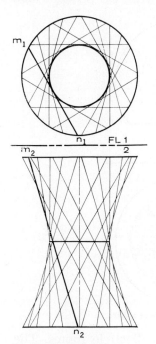

FIG. 6.20. Hyperboloid of revolution of one sheet.

Any plane section of this surface which contains the axis will be a hyperbola, and any plane section which is at right angles to the axis will be a circle. The smallest possible circular section is called "the circle of the gorge," and it will have a radius equal to the shortest distance between the two lines.

This surface is usually represented by showing the axis, several positions of the line which revolves, the two end or base circles, and the circle of the gorge. Figure 6.20 shows two views of a hyperboloid of revolution which is generated by revolving the line MN around the vertical axis.

This surface cannot be developed.

The best-known application of this hyperboloid is in the design of hypoid gears. Each tooth is straight, occupying a position of the generatrix. If the elevation view in Fig. 6.20 were cut off at about one-eighth height, the remaining lower portion could be visualized as a hypoid gear, similar in appearance to a bevel gear.

6.57   Practice problems

See Chapter 8, Group 62.

6.58   Sphere

A sphere is a double-curved surface that is generated by revolving a circle about a line passing through its center. A sphere will appear exactly the same in all orthographic views. It always appears to be the true size of the circle by which it is generated. It is therefore represented in any view by merely showing its generating circle.

Any plane section through a sphere at any angle will be a circle. If the cutting plane contains the center of the sphere, the section is called a "great circle"; if it does not contain the center of the sphere, the section is said to be a "small circle."

Figure 6.21 shows two views of a sphere which is cut by a plane passing through its center. The circle cut out is a great circle which appears as an ellipse in the front view. Figure 6.22 shows the same sphere cut by any plane which does not contain its center. The circle cut by this plane is seen to be a small circle, and its center lies on a line which is perpendicular to the cutting plane and passes through the center of the sphere.

The earth is practically a sphere, and those who study naviga-

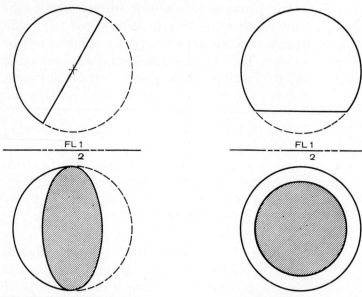

FIG. 6.21. Great circle of sphere.   FIG. 6.22. Small circle of sphere.

tion must be familiar with great and small circle routes. The domes of some buildings are portions of spheres, and many bearings depend upon spherical balls for their operation. Also some steel water tanks are made with a hemispherical bottom, and gas retorts often have a top which is partly spherical.

6.59   To locate a point on a sphere, having given one view

If one view of a point is given as being on the surface of a sphere, a small-circle element may be drawn containing the point. This element should be drawn so that it is a straight line in one view and a circle in the other. The point will have to be on this element in both views. A large-circle element could also be used, but might require an extra view to show it as a circle. The construction is not shown here.

6.60   To find where a line pierces a sphere

A plane is drawn so as to contain the given line and either a great or a small circle on the sphere. A view showing the true size of that plane will show the two points where the given line intersects the circle which is cut out by the plane. These two points are on both the line and the surface of the sphere, and they are therefore the desired piercing points. The solution is not shown.

6.61   To draw a plane tangent to a sphere and containing a given line

A new view is drawn showing the sphere and also showing the line appearing as a point. Any plane containing the given line will show as an edge in this view. The required plane is so drawn that it contains the line (which is a point in this view) and consequently is tangent to the sphere, on either side. Two solutions are possible. The tangent plane may be located in any other views by selecting any two random lines upon it and projecting them to the views desired. The solution is not shown.

6.62   **Practice problems**

See Chapter 8, Group 63.

6.63   **Approximate development of a sphere**

The sphere itself is an undevelopable surface. It is sometimes necessary to construct a spherical surface from steel plate or other material, and then an approximate method must be used. Two methods are in use, both of which consist in breaking up the surface of the sphere into small portions and assuming each portion to be a part of a surface which can be developed.

The meridian method, shown in Fig. 6.23, consists in cutting up the surface by a series of planes passing through the center of the sphere and assuming that each small section is a portion of a cylinder. This small portion is easily developed, as is shown in the drawing. If the planes are spaced an equal distance, or angle, apart around the sphere, it is necessary to develop only one piece, for all the cut sections are alike. This is known in practice as the orange-peel method.

The zone method, shown in Fig. 6.24, consists in cutting the sphere by a series of parallel planes and assuming that each section of the surface cut is a portion of a cone of revolution. Each of these sections has to be developed independently, for they all lie on cones of different size, that is, for half a sphere. Two of the cones are shown in the drawing, and their development is also shown.

In using either of these methods the surface should be divided up into as many sections as possible in order to approach more closely the actual spherical surface.

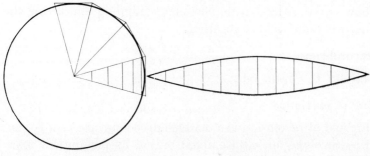

FIG. 6.23. Meridian method of developing a sphere.

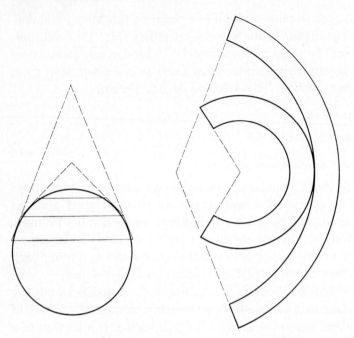

FIG. 6.24. Zone method of developing a sphere.

6.64   Torus or annulus

The torus is a double-curved surface generated by revolving a circle about a straight-line axis which does not contain the center of the circle. If this axis does not intersect the circle, an open torus is generated with a hole through its center. A torus is represented by showing just the extreme projecting elements in each view (see Fig. 6.25).

Elbows furnish nearly all the practical examples of a torus. An elbow is usually a quarter of an approximate torus. The shop method of making one of these steel elbows is to make a series of diagonal cuts, by means of a torch, across a stock steel pipe. Adjacent cuts are reversed in direction, thus giving trapezoidal-shaped sections of pipe. Every other section of pipe is then reversed, the long sides all being turned on the same side to make the elbow curve. After a little necessary chipping and fitting the sections are then welded together.

6.65   Practice problems

See Chapter 8, Group 64.

6.66   Ellipsoid of revolution

The ellipsoid of revolution is a double-curved surface that is generated by revolving an ellipse about one of its rectangular axes. If it is revolved about its major axis, the surface generated is a prolate ellipsoid. It has the appearance of a football. If the ellipse is revolved about its minor axis the surface generated is an oblate ellipsoid. It has the appearance of a discus or a door knob. Tanks of this shape are now used for large storage tanks for gas.

A prolate ellipsoid will appear as an ellipse in two views and in the third view as a circle having a diameter equal to the minor axis.

An oblate ellipsoid will appear as an ellipse in two views and in the third view as a circle having a diameter equal to the major axis.

6.67   Paraboloid of revolution

The paraboloid of revolution is a double-curved surface that is generated by revolving a parabola about the symmetrical axis which passes through its focus. In a three-view drawing of this surface, two views will show it as a parabola, and the third view will show it as a circle.

This surface occurs often in searchlight reflectors, in modified form, because of its peculiar properties in reflecting parallel rays. If it is made out of sheet metal it must be pressed into shape, for it is undevelopable.

6.68   Hyperboloid of revolution of two sheets

The hyperboloid of revolution of two sheets is a double-curved surface which is generated by revolving a hyperbola about an axis containing both its foci. It must be remembered that the

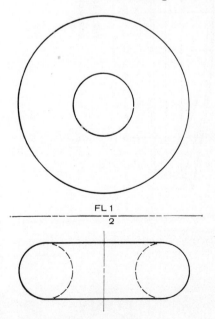

FL 1
––––
2

FIG. 6.25. Torus or annulus.

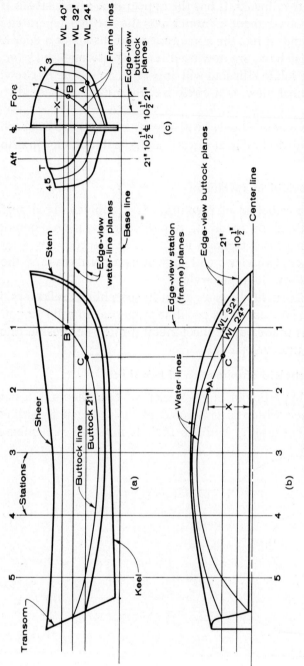

FIG. 6.26. Fairing a boat hull. (a) Sheer plan. (b) Half-breadth plan. (c) Body plan. Note: Buttock 10½ in. and water line 40 in. have not been drawn in the sheer and half-breadth plans.

mathematical equation for a hyperbola places the curve into opposite quadrants and that the two curves are entirely separate. Therefore this surface is said to have two sheets, because it is in two separate and distinct pieces. It must be distinguished from the hyperboloid of revolution of one sheet in Section 6.56. In a three-view drawing of this surface, two views would show it as two separate hyperbolas and the third view would show it as a circle.

6.69   Practice problems

See Chapter 8, Group 65.

6.70   Fairing double-curved surfaces

Complex double-curved surfaces such as those occurring in boat hulls, airplane fuselages, and automobile bodies are usually described by drawings showing contours cut out by three sets of mutually perpendicular planes. A well-designed surface is said to be *fair* when it is continuous and smooth with no abrupt bulges or indentations except where intended. If the surface itself is fair, then the contours will be continuous and smooth, not conflicting with each other. The design of a fair surface is really a cut-and-try process, with the success and speed of the operation depending on the experience of the designer and his instinctive appreciation of a fair line.

An example of a faired surface is shown in the drawing of the boat in Fig. 6.26. The objective here is to get a set of frame lines in the body plan from which temporary molds or patterns can be made which, when erected at station points on the keel and held in place by temporary ribbands or strips of wood running longitudinally from stem to stern, form the hull shape over which permanent frames of oak can be steamed and bent.

The frame lines, when projected by the principles of orthographic projection, must produce contours, in this case water lines and buttock lines, that are fair. If these lines are not fair, the frame lines must be adjusted until the resulting contours are pleasing to the eye. The final step on the drawing board is to scale offsets from the body plan so that the frame lines can be laid out full size in the ship loft. In the old days the same result was achieved, but more crudely, by making a wooden half model of the hull to a reduced scale and measuring offsets at stations for the frame lines.

Referring to Fig. 6.26, point A on frame 2 and WL 24 in. in the half-breadth plan is located by laying off distance X taken from the body plan at the intersection of frame 2 and WL 24 in. The same procedure is followed with the intersections of the remaining frame lines and WL 24 in. in the body plan to produce the complete WL 24 in. in the half-breadth plan. To find buttock line 21 in. in the sheer plan it is necessary to project horizontally from the body plan to the proper station in the sheer plan the intersection of each frame line with buttock 21 in. as illustrated by point B. A further and important test of projection is shown at point C. Point C in the sheer plan is the intersection of WL 24 in. and buttock 21 in., and C in the half-breadth plan is also the intersection of WL 24 in. and buttock 21 in. By the principles of orthographic projection, these points should line up vertically. All such intersections should be tested for accuracy.

Notice that WL 40 in. and buttock 10½ in. have been omitted in the sheer and half-breadth plans to avoid complicating the drawing. These would ordinarily be shown.

6.71   Miscellaneous

Most of the double-curved surfaces that have been introduced here have been surfaces of revolution, because these occur more frequently and because their generation can be more easily defined. A countless number of other double-curved surfaces might be generated by complex motions of curved lines. Most of these surfaces do not even have a name and seldom occur in engineering practice. They have been omitted for this reason.

A sufficient number of surfaces have been explained to enable the reader to classify practically every surface with which he will ever come in contact in actual experience. He would know what kind of elements were on the surface, he could represent the surface, solve whatever problems were necessary in connection with it, and make an approximate development of it if necessary. A few of the more common cases of intersections of the various surfaces of this chapter will be considered in the chapter following.

7

INTERSECTION OF SURFACES

7.1   Introduction

Occasionally in engineering construction, and often in large special pipe fittings, an object is made up of two different surfaces that intersect each other, each surface extending only to its line of intersection. For development work this line of intersection must be accurately determined, which means that it will have to be located accurately in two orthographic views. A large variety of problems might be introduced in this chapter, corresponding to all the different possible combinations of the surfaces which were discussed in Chapter 6. However, it is intended to give only a very few representative problems which may occur most frequently in engineering practice. In these problems methods of solution will be introduced which are quite typical as far as the space analysis is concerned. A general procedure will then be given containing a few rules and suggestions to be followed in the solution of miscellaneous intersection problems.

143

7.2   Two plane surfaces: prisms and pyramids

The intersection of objects which are bounded entirely by plane surfaces may be determined completely by the methods of intersection of planes or a line piercing a plane. These methods have already been explained in Sections 3.21 and 3.25.

7.3   Plane surface and any other surface

First method

A new view of both surfaces in which the plane appears as an edge will show where each element of the other surface intersects the plane, whether the elements used are straight lines or curved. Each intersection point may then be projected to the proper element in the desired views.

Second method

If the nonplanar surface has straight-line elements, the points where these elements pierce the plane may be found by using the method of two views only, Section 3.21. A series of these points, if connected with a fair curve, will give the entire line of intersection.

Third method

If the nonplanar surface has circular elements for its simplest ones, the better method is to intersect both the given surfaces with a third plane, which will cut out a circular element and a straight-line element from the two given surfaces. It will be necessary to use a series of these planes in order to determine a sufficient number of points.

7.4   Two cylinders with their bases in the same plane

First method

The points at which the straight-line elements on one cylinder pierce the other cylinder are points on both cylinders; that is, on the line of intersection of the two cylinders. These points may be determined by using either method of Section 6.11.

Second method

A series of planes could be passed cutting circular or elliptical elements on both cylinders. Since these two elements would lie in the same plane, they must intersect and give points on the required line of intersection or miss each other altogether. This method should be used only when circular elements are cut from both cylinders.

Third method

Analysis.   A plane may be passed so that it is parallel to both cylinders and therefore will cut straight-line elements on both. These elements, two on each cylinder, lie in the same plane, and they must intersect (or be parallel), giving four points on the line of intersection. Other planes, all parallel to the first one, are passed to determine other points on the desired curve.

Explanation (see Fig. 7.1). The two oblique cylinders are given in the position shown. From any random point, such as X, two lines are drawn parallel to these two cylinders. These lines determine a plane which is parallel to both cylinders and which will cut straight-line elements on both (if it intersects them). This plane intersects the plane of the bases in the line MN, which line determines the two elements the plane cuts out of each cylinder. These four elements are shown in both views; the four points at which they intersect are on the line of intersection of the two cylinders. These points are determined independently in each view, and they should be checked by projection between views. A series of planes taken parallel to this first plane will establish sufficient points to determine the entire line of intersection.

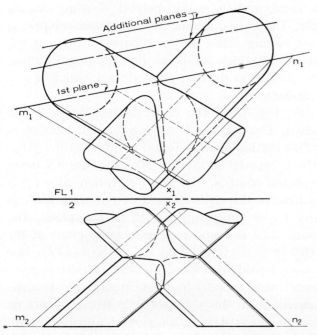

FIG. 7.1. Two cylinders. Bases in the same plane.

These cutting planes, although they are taken at random, should be drawn first, so that they will contain the extreme elements on the cylinders in both views. This may require eight planes, and they will determine the exact tangent points at which the curve touches all the extreme elements. After these eight tangent points have been determined, other planes are taken where they are needed to complete the curve. In order for a point on a line of intersection to be visible, it must lie on a visible element on both surfaces. This method may also be used in finding the line of intersection of two prisms, or a prism and a cylinder.

Note: Do not develop the surfaces from these views. If the development is required use the third method of Section 7.5.

7.5   Two cylinders with their bases in different and nonparallel planes

First method

The first method is the same as the first method of Section 7.4. It applies equally well regardless of the position of the bases.

Second method

Analysis.   The axes of the two cylinders would not occupy any different relative positions in space from those they would for the third method of Section 7.4. The cylinders themselves are just cut off by different planes. Therefore the space analysis remains just the same, and the same method of using cutting planes will apply. The only difference in the detail solution is that the assumed cutting plane will now have to intersect the two different planes in which the bases lie. Additional cutting planes will all be parallel to the first plane drawn, which is parallel to both cylinders.

Explanation (see Fig. 7.2). The two cylinders are given in the position shown. The plane *AXZ* is determined, parallel to both cylinders. This cutting plane is found to intersect the plane of the base of the *B* cylinder in the line *MN*, the line *RN* being used to determine the point *N*. The front elevation shows *MN* intersecting the base of this cylinder and determining the two elements of the cylinder which are cut out by this plane. The same cutting plane *AXZ* is found to intersect the plane of the base of the *A* cylinder in the line *AK*, by use of the auxiliary line *XY* on the plane. In the side elevation this line *AK* is seen intersecting the base of the *A* cylinder and determining the two elements which are cut from this cylinder. The two elements on each cylinder are now located in all the views by projection. All four elements lie in the plane *AXZ*, and therefore they intersect

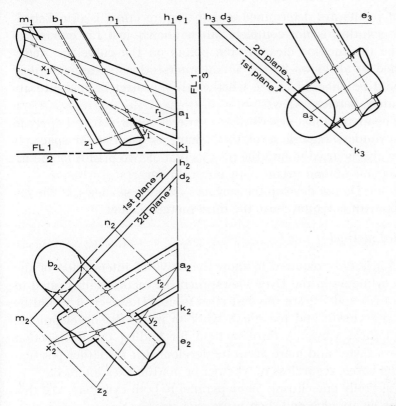

FIG. 7.2. Two cylinders. Bases in different and nonparallel planes. Showing the method of locating four points only. The solution is not complete.

at the points shown, giving four points on the curve of intersection.

The two planes containing the cylinder bases form a dihedral angle with its vertex in the vertical line *HE*. Any plane cutting across this dihedral angle must cut out a plane angle with its vertex on the line *HE*. In other words, the two intersection lines *MN* and *AK* are the sides of this plane angle, and they must meet on the line *HE*. They are seen to check by meeting at the point *H*. This fact is used in locating additional cutting planes. The second plane is now assumed, and in the front elevation its intersection is drawn at random, either tangent to or intersecting the base of the *B* cylinder. This intersection with the base plane meets the line *HE* at *D*. The point *D* is then located on *HE* in the side elevation, giving d_3. From d_3 the intersection with the plane of the base of the *A* cylinder is drawn parallel to the intersection of the first plane. In Fig. 7.2 the second plane is drawn tangent to the *B* cylinder, to show that it is the lowest plane that may be used.

Other planes are drawn in this same way until the special tangent points are determined and the entire curve is established. The solution is not complete but is shown just far enough to make the method clear. If the points on the curve are determined in all views by the actual intersection of elements they should be checked to see whether they project or measure accurately between views.

The method just described is a general one which will apply to all cylinders regardless of their shape or position in space. It may also be used to find the intersection of two prisms or the intersection of one prism with one cylinder.

Note: Do not develop the surfaces from these views. If the development is required use the third method below.

Third method

If it is only required to show the curve of intersection of the two cylinders in the given views, then the methods illustrated in Figs. 7.1 and 7.2 are the best ones to use. However, it would be too inaccurate and too much work to make the developments from these views. A third method for finding the intersection allows faster and more accurate development for either position of the bases, regardless of whether or not the axes intersect.

Pass only one cutting plane parallel to both cylinders. See this plane as an edge and then in its true size. In this true-size view both cylinder axes will be in their true lengths.

Temporarily call this true size view a *plan* and intersect both cylinders with a series of level planes. The curve of intersection will be easy to obtain and very much simplified. Since both axes are in their true lengths, both cylinders may be developed directly from this view. The solution is not shown.

7.6   Practice problems

See Chapter 8, Group 66.

7.7   Two cones with their bases in the same plane

First method

By the method of Section 6.31, the two points at which any straight-line element on one cone pierces the other cone may be determined. The entire curve of intersection may be found by repeating this process with a sufficient number of elements. This is a rather laborious method, and it does not establish special points on the curve with sufficient accuracy.

Second method

Both cones may be cut by a series of planes parallel to their bases. The elements cut from the cones will probably be circles or ellipses. The intersections of these elements which lie in the same plane will be points on the desired line of intersection. The only condition which would make it practicable to use this method would occur when the planes cut circular elements on both cones. Even then the special tangent points on the curve are found only by a cut-and-try method.

Third method

Analysis.   A cone has straight-line elements, all of which pass through its vertex. Therefore every cutting plane which cuts straight-line elements on the cone must contain the vertex of the cone. Also, for the same cutting plane to cut straight-line elements on two cones, it must contain the vertices of both cones. It must also contain a straight line connecting these two vertices. If this straight line is drawn and produced until it pierces the plane containing the bases of both cones, this piercing point will also have to lie in every cutting plane. If all the cutting planes are produced until they intersect the plane of the bases, these intersections will all meet at this piercing point of the line of the vertices with the plane of the bases. These intersections may be drawn in at random, so long as they cut across the bases of both cones, and the points where they intersect the bases determine the elements which will intersect each other.

Explanation (see Fig. 7.3). The two oblique cones are given as shown. The line *AC*, which contains both vertices, pierces the plane of the bases at the point *N*. The first cutting plane is drawn at random, its intersection with the plane of the bases being the line *MN*. This line is seen to cut across both bases, thus determining two elements on each cone. These four elements, which lie in the same cutting plane, intersect as shown, and thus determine four points on the curve of intersection. The four elements are shown in both views. Additional planes are drawn at random, so that their intersections with the base plane cut across both bases and pass through the point *N*. The point *N* is the controlling point for the entire solution. The first planes used should always contain the extreme elements on both cones for both views, in order to locate the exact points at which the curve comes tangent to them.

This method may also be used in finding the line of intersection of two pyramids, or a pyramid and a cone.

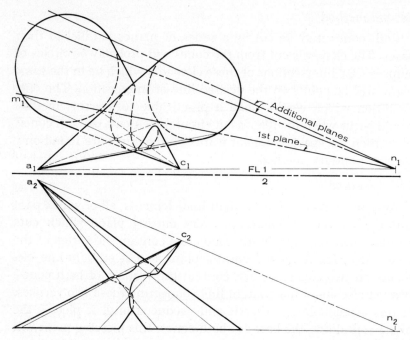

FIG. 7.3. Two cones. Bases in same plane.

7.8 Two cones with their bases in different and nonparallel planes

First method

The first method of Section 7.7 will apply just the same, even though the bases are not in the same plane.

Second method

Analysis. A cutting plane is used which contains the line connecting the vertices of the two cones. The intersection of this plane with each of the two base planes will have to contain the point at which the line of the vertices pierces these planes. These two lines of intersection with the base planes may then be drawn in at random, one through each piercing point, provided they meet on the line of intersection of the two base planes and also provided that each line cuts across a base of a cone. The points at which these lines of intersection intersect the bases will determine two straight-line elements on each cone. These four elements will intersect to give four points on the intersection of the cones. Additional planes will give more points, but no two of these planes are parallel, since each must contain the line of the vertices.

Explanation (see Fig. 7.4). The two cones are in the positions shown with their vertices at *A* and *B*. In the front elevation, the line of the vertices, *AB*, is seen to pierce the plane of the base of the A cone at N and of the B cone at M. The point M is located in the plan and the point N is located in the side elevation. These two points are now the controlling points for the solution. In the plan, the intersection of the first cutting plane is drawn at random from M across the base of the B cone. This intersection meets the line of intersection of the two base planes at the point H. H is projected to the side elevation, and the intersection of this cutting plane with the plane of the base of the A cone must be the line NH. In the plan the two elements of the B cone are determined and in the side elevation the two elements on the A cone are determined. All four elements lie in the first cutting plane and they are all located in every view. The four points where they intersect are apparent. These points should check between views by projection and by measurement.

The solution is not complete in Fig. 7.4. A second cutting plane is shown; it intersects the two base planes in the lines MK and NK. Any other planes may be drawn in the same manner.

The solution by this second method will apply for all problems involving the intersection of two cones, two pyramids, or a pyramid with a cone.

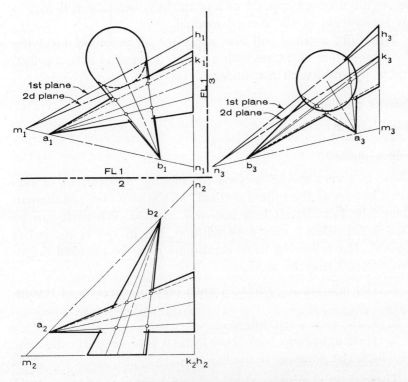

FIG. 7.4. Two cones. Bases in different and nonparallel planes. Showing the method of locating four points only. The solution is not complete.

7.9   Practice problems

See Chapter 8, Group 66.

7.10   Cone and cylinder

First method

By the piercing-point method and using two views only, the points at which any straight-line element on either surface pierces the other surface may easily be determined.

Second method

A new view may be drawn in which the cylinder appears as an edge. In this view the point at which any straight-line element on the cone intersects the cylinder will be apparent.

Third method

A line may be drawn through the vertex of the cone and parallel to the cylinder. A series of planes are taken so that they contain this line and cut both surfaces. These planes will cut straight-line elements on both surfaces, which will have to intersect. This solution is performed more easily in the view in which the line through the vertex appears as a point, in which case it is merely an application of the second method.

This third method will also apply for all problems involving the intersection of a cone with a prism, a pyramid with a prism, and a pyramid with a cylinder.

7.11   Practice problems

See Chapter 8, Group 66.

7.12   Sphere method

There is a very special method for solving intersections of surfaces known as the sphere method, with which every draftsman should be familiar. It is a one-view method. Where it can be used it furnishes a very easy solution, but its use is somewhat limited. The following three conditions must be satisfied before this method may be used:

1. The intersecting surfaces must both be surfaces of revolution.

2. Their axes must intersect.

3. Their axes must both show in their true length in the view in which the solution is made.

These three conditions should always be checked before use of this method is attempted.

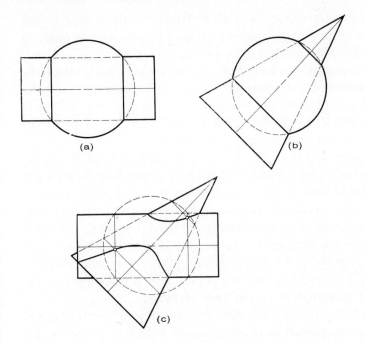

FIG. 7.5. Sphere method.

Analysis.   If the three conditions given have been satisfied, the intersection of any surface of revolution with a sphere is a circle which appears as an edge in the view in which the solution is made. Figure 7.5*a* and *b* shows a cylinder and a cone, respectively, intersecting a sphere, and the intersections, in each case, are seen to be straight lines.

At the point of intersection of the axes of the two intersecting surfaces a sphere is drawn, having any diameter. It intersects both of the given surfaces in circles which appear as straight lines. Both of these circles also lie on the surface of the sphere. If they intersect on the sphere, that point of intersection also lies on both of the given surfaces and is therefore a point on their line of intersection. A series of different-sized spheres is drawn until the curve of intersection is entirely established. The solution is given in Fig. 7.5*c*, where only one sphere is shown.

This method may be used with any combination of two surfaces of revolution provided the three conditions are satisfied. Since many actual metal pieces are surfaces of revolution with intersecting axes, this method has considerable practical value.

7.13   General procedure

Each combination of two intersecting surfaces presents a situation a little different, and requires careful analysis before the solution is attempted. There is no one method by which all intersection problems may be solved. However, there is a rather general procedure of analysis which may be followed in order to arrive at the best method of solution.

The first step in this analytical procedure is to have clearly in mind the nature of each surface, the way it is generated, and the different kinds of elements which may lie upon it. If a surface is double-ruled, clear determination should be made as to which way both sets of straight-line elements lie.

The second step is to think of cutting across both the given surfaces with a plane which will contain some element on each surface. It will have to be decided which way this plane must be passed in order to cut out the simplest elements on both surfaces. Straight-line elements should always be used if possible. Circular elements are the next easiest to handle. If elements on the two given surfaces lie in this same cutting plane, they will intersect (in general) at one, two, or four points, which will be points on the line of intersection of the two surfaces.

The third step in the analysis is to determine which way other planes must be passed in order to locate sufficient points to determine the entire line of intersection. Sometimes the cutting planes are all parallel, and sometimes they are intersecting, according to the nature of the given surfaces. A sufficient number of additional cutting planes should be passed to determine completely and accurately the desired line of intersection.

Some of the more unusual intersections are now listed with a brief statement of the method recommended for the solution of each one. Each method is selected by analysis closely following the general procedure explained in this section.

1. *Two spheres.* Use a series of planes, preferably horizontal or vertical, any one of which cuts a circle from each sphere. Solve in the view where these circles appear as circles.

2. *Prism and sphere.* Use a series of planes parallel to the prism. Vertical planes give the easiest solution.

3. *Pyramid and sphere.* Use a series of planes containing the vertex of the pyramid. Vertical planes are preferable.

4. *Cylinder and sphere.* Same as for prism and sphere.

5. *Cone and sphere.* Same as for pyramid and sphere.

6. *Prism and hyperbolic paraboloid.* Use a series of planes containing straight-line elements on the warped surface and cutting straight lines on the prism surfaces.

7. *Pyramid and hyperbolic paraboloid.* Same as for prism and hyperbolic paraboloid.

8. *Cylinder and hyperbolic paraboloid.* Use a series of planes containing straight-line elements on the warped surface. These planes will cut ellipses, circles, or straight lines on the cylinder, depending upon its position relative to the warped surface. Or a series of planes could be used cutting straight-line elements on the cylinder and curved lines on the warped surface.

9. *Cone and hyperbolic paraboloid.* Use a series of planes containing the vertex of the cone and cutting curved lines on the warped surface.

10. *Torus and cylinder.* Use a series of planes perpendicular to the axis of the torus. These planes will cut circles on the torus and ellipses, circles, or straight lines on the cylinder.

11. *Torus and cone.* Use a series of planes perpendicular to the axis of the torus and cutting a series of ellipses or circles on the cone.

7.14   Miscellaneous practice problems

See Chapter 8, Group 66.

8

PRACTICE PROBLEMS

The problems contained in this chapter are given for the purpose of furnishing practice in the use of each principle which has been explained. As soon as a principle has been studied, it should be tried out by solving some of the various problems listed under that group. These problems are all given without any data, but they should be laid out approximately in the proportion and in the position as given. In a few problems the proportion between the width and height of a solid object will be suggested. All the problems in this chapter should be solved by *freehand,* using only a pencil.

These problems are designed to test a student's understanding of a theoretical method and his ability to apply that method in obtaining a solution in a large variety of situations.

For convenience each problem has been given a classification number, such as 8.24.7. This notation means that this problem is number 7 in Group 24 in Chapter 8.

Group 1. Three or more ordinary views of an object

In each problem draw the specified views of the object and show all hidden lines in all views. The rule in shop drawing is to omit dash lines unless they make the drawing clearer. Since the chief purpose of these problems is to give practice, the locating of the dash lines will give just that much more practice. Also show and label all the folding lines.

8.1.1. Draw the plan, the front elevation, and the right side elevation of the object in Fig. 8.1(1).

8.1.2. Draw the plan, the front elevation, and both side elevations of the object in Fig. 8.1(2).

8.1.3. Draw the plan, the front elevation, and both side elevations of the object in Fig. 8.1(3).

8.1.4. Draw the two given views and one side elevation of the object in Fig. 8.1(4).

8.1.5. Draw the two given views and both side elevations of the object in Fig. 8.1(5).

8.1.6. Draw the two given views and both side elevations of the object in Fig. 8.1(6).

8.1.7. Draw the two given views and the right side elevation of the object in Fig. 8.1(7).

8.1.8. Draw the two given views and the left side elevation of the object in Fig. 8.1(8).

8.1.9. Draw the plan and both side elevations and complete the front elevation of the object in Fig. 8.1(9).

8.1.10. Draw the plan, the right side elevation, and the front elevation of the object in Fig. 8.1(10).

8.1.11. Draw the two given views and the left side elevation of the object in Fig. 8.1(11).

8.1.12. Draw the given views and the left side elevation of the object in Fig. 8.1(12).

Group 2. Auxiliary elevation views of an object

In each problem draw the two given views and the auxiliary elevation looking in the direction indicated by the arrow marked C. Show the entire object, including all dash lines, in every view.

8.2.1. Object in Fig. 8.2(1).
8.2.2. Object in Fig. 8.1(4).
8.2.3. Object in Fig. 8.1(5).
8.2.4. Object in Fig. 8.1(6).
8.2.5. Object in Fig. 8.1(7).
8.2.6. Object in Fig. 8.1(8).
8.2.7. Object in Fig. 8.1(9) (omit the front elevation).
8.2.8. Object in Fig. 8.1(10) (omit the front elevation).
8.2.9. Object in Fig. 8.1(11).
8.2.10. Object in Fig. 8.1(12).

Group 3. Inclined views taken from the front elevation

In each problem draw the two given views and an inclined view looking in the direction indicated by the arrow marked A. Show the entire object, including all dash lines, in every view.

8.3.1. Object in Fig. 8.2(1).
8.3.2. Object in Fig. 8.2(2).
8.3.3. Object in Fig. 8.2(3).
8.3.4. Object in Fig. 8.2(4). Show only the lower part of the object as cut by the plane AA.
8.3.5. Object in Fig. 8.2(5).
8.3.6. Object in Fig. 8.2(6).
8.3.7. Object in Fig. 8.1(6).
8.3.8. Object in Fig. 8.1(7).
8.3.9. Object in Fig. 8.1(11).
8.3.10. Object in Fig. 8.1(12).

Group 4. Inclined views taken from auxiliary elevations or other inclined views

In each problem draw the two given views and the two extra views specified. Show the entire object, including all the dash lines, in every view.

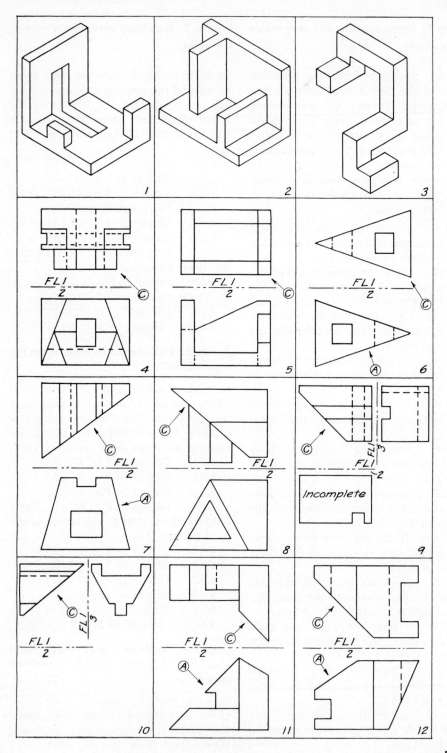

FIG. 8.7

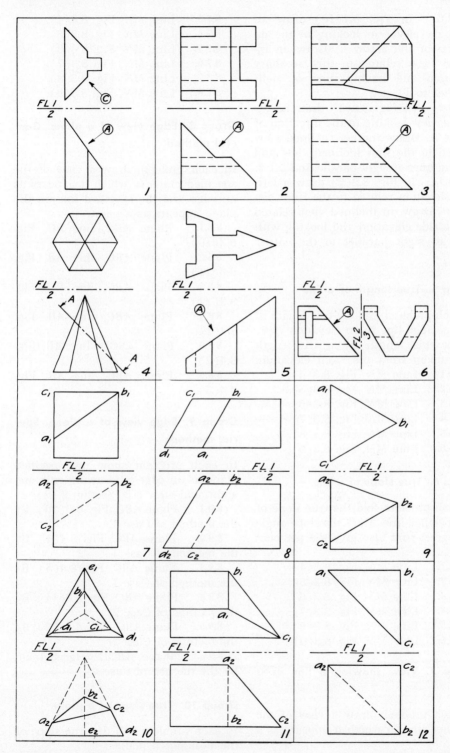

FIG. 8.2

8.4.1.   In Fig. 8.1(10) draw an auxiliary elevation looking in the direction of the arrow C. Draw an inclined view related to this auxiliary elevation and having lines of sight parallel to FL 1-2.

8.4.2.   In Fig. 8.2(1) draw an inclined view looking in the direction of arrow A. Draw another inclined view related to the first inclined view and having lines of sight parallel to FL 1-2.

8.4.3.   In Fig. 8.2(7) draw a left-side elevation related to the front elevation. Draw an inclined view related to the side elevation and looking with lines of sight parallel to the corner C-A.

Group 5. True length of a line

In each problem find the true length of the specified line in an elevation view. Check this length by a true-length view taken from the front elevation.

8.5.1.   Line MN, Fig. 8.3(4).
8.5.2.   Line MN, Fig. 8.3(5).
8.5.3.   Line MN, Fig. 8.3(6).
8.5.4.   Line MN, Fig. 8.3(7).
8.5.5.   Line MN, Fig. 8.3(8).
8.5.6.   Line MN, Fig. 8.3(9).

Group 6. True slope of a line

In each problem find the true slope of the specified line. Mark the slope angle in degrees and also give the per cent grade.

8.6.1.   Line MN, Fig. 8.3(4).
8.6.2.   Line MN, Fig. 8.3(5).
8.6.3.   Line MN, Fig. 8.3(6).
8.6.4.   Line MN, Fig. 8.3(7).
8.6.5.   Line MN, Fig. 8.3(8).
8.6.6.   Line MN, Fig. 8.3(9).

Group 7. View showing a line as a point

In each problem draw a view of the specified line in which it appears as a point.

8.7.1.   Line MN, Fig. 8.3(2).
8.7.2.   Line MN, Fig. 8.3(3).

8.7.3.   Line MN, Fig. 8.3(4).
8.7.4.   Line MN, Fig. 8.3(5).
8.7.5.   Line MN, Fig. 8.3(6).
8.7.6.   Line MN, Fig. 8.3(7).
8.7.7.   Line MN, Fig. 8.3(8).
8.7.8.   Line MN, Fig. 8.3(9).

Group 8. Edge view of a plane. General method

In each problem draw a view of the specified plane in which it appears as an edge and the specified line on the plane appears as a point.

8.8.1.   Plane ABC, line BC, Fig. 8.3(10).
8.8.2.   Plane ABC, line AB, Fig. 8.3(11).
8.8.3.   Plane ABC, line BC, Fig. 8.3(12).
8.8.4.   Plane ABC, line AB, Fig. 8.4(1).
8.8.5.   Plane ABCD, line AB, Fig. 8.4(2).
8.8.6.   Plane ABC, line AB, Fig. 8.4(5).

Group 9. Edge view of a plane. Special methods

In each problem show the specified plane as an edge by drawing only one additional view (see Section 2.12).

8.9.1.   Plane ABC, Fig. 8.3(10). By the method of Case 1.
8.9.2.   Plane ABC, Fig. 8.4(1). By the method of Case 1.
8.9.3.   Plane ABC, Fig. 8.4(5). By the method of Case 1.
8.9.4.   Plane ABC, Fig. 8.3(11). By the method of Case 2.
8.9.5.   Plane ABC, Fig. 8.3(12). By the method of Case 2.
8.9.6.   Plane ABCD, Fig. 8.4(2). By the method of Case 2.

Group 10. True slope of a plane

In each problem find the true slope of the designated plane.

8.10.1.   Plane ABC, Fig. 8.3(10).
8.10.2.   Plane ABC, Fig. 8.3(11).

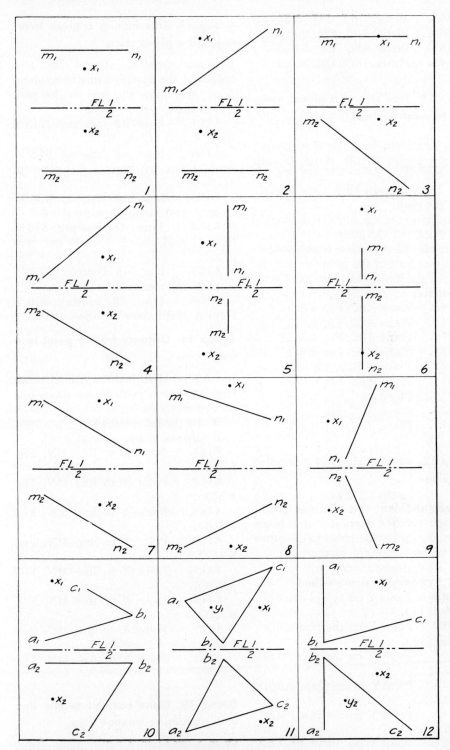

FIG. 8.3

8.10.3.   Plane *ABC*, Fig. 8.3(12).
8.10.4.   Plane *ABC*, Fig. 8.4(1).
8.10.5.   Plane *ABC*, Fig. 8.4(3).
8.10.6.   Plane *ABC*, Fig. 8.4(4).

Group 11. True size of a plane, using only two extra views

In each problem draw the view looking at right angles to the plane, by two methods.

I. By taking an edge view directly from the plan.

II. By taking an edge view directly from the front elevation.

The size of the plane is assumed to mean that part of the plane within the limits of the lines which are used to fix it in space.

8.11.1.   Plane *ABC*, Fig. 8.3(10).
8.11.2.   Plane *ABC*, Fig. 8.3(11).
8.11.3.   Plane *ABC*, Fig. 8.3(12).
8.11.4.   Plane *ABC*, Fig. 8.4(1).
8.11.5.   Plane *ABCD*, Fig. 8.4(2).
8.11.6.   Plane *ABC*, Fig. 8.4(3).
8.11.7.   Plane *ABC*, Fig. 8.4(4).
8.11.8.   Plane *ABC*, Fig. 8.4(5).
8.11.9.   Plane *ABC*, Fig. 8.4(6).

Group 12. Various views of any point in a plane

In each problem the plane is given, and one view of a point which is on that plane. Show the point in the other view by two different methods.

I. By using two views only.

II. By the edge-view method.

8.12.1.   Point *Y* on the plane *ABC*, Fig. 8.3(11).

8.12.2.   Point *Y* on the plane *ABC*, Fig. 8.3(12).

8.12.3.   Point *Y* on the plane *ABC*, Fig. 8.4(1).

8.12.4.   Point *Y* on the plane *ABCD*, Fig. 8.4(2).

8.12.5.   Point *Y* on the plane *ABC*, Fig. 8.4(3).

8.12.6.   Point *Y* on the plane *ABC*, Fig. 8.4(4).

Group 13. Line having a given bearing and a given slope

In each problem assume the true length of the specified line to be about 3 in. and show the line in the plan and front elevation views.

8.13.1.   Line *AB* bearing N45°E from *A* and falling 30°.

8.13.2.   Line *AB* bearing S15°W from *A* and rising on a 100 per cent grade.

8.13.3.   Line *AB* bearing N30°W from *A* and having a slope = −0.5.

8.13.4.   Line *AB* bearing S75°E from *A* and rising on a 45 per cent grade.

8.13.5.   Line *AB* bearing N75°W from *A* and rising 60°.

8.13.6.   Line *AB* bearing S30°E from *A* and having a slope = +⁷⁄₁₀.

Group 14. Distance from a point to a line

In each problem find the true distance from the given point to the given line by two methods.

I. By the line method.

II. By the plane method.

8.14.1.   Point *X* to line *MN*, Fig. 8.3(4).

8.14.2.   Point *X* to line *MN*, Fig. 8.3(5).

8.14.3.   Point *X* to line *MN*, Fig. 8.3(6).

8.14.4.   Point *X* to line *MN*, Fig. 8.3(7).

8.14.5.   Point *X* to line *MN*, Fig. 8.3(8).

8.14.6.   Point *X* to line *MN*, Fig. 8.3(9).

8.14.7.   Point *X* to line *AC*, Fig. 8.3(11).

8.14.8.   Point *X* to line *MN*, Fig. 8.3(2).

Group 15. Plane parallel to one line and containing another line

In each problem draw a plane which contains the line *AB* and is parallel to

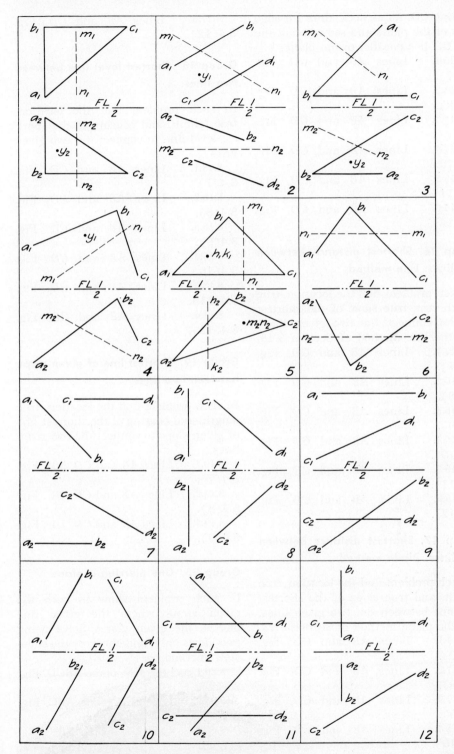

FIG. 8·4

the line CD. As a check, draw an edge view of the plane and see whether the line CD lies parallel to the plane.

8.15.1. Lines AB and CD, Fig. 8.4(7).

8.15.2. Lines AB and CD, Fig. 8.4(8).

8.15.3. Lines AB and CD, Fig. 8.4(9).

8.15.4. Lines AB and CD, Fig. 8.4(10).

8.15.5. Lines AB and CD, Fig. 8.4(11).

8.15.6. Lines AB and CD, Fig. 8.4(12).

Group 16. Shortest distance between two lines. Line method

In each problem find the location, true length, and true slope of the shortest line to connect the two given lines. Use the line method of Section 3.13.

8.16.1. Lines AB and CD, Fig. 8.4(7).

8.16.2. Lines AB and CD, Fig. 8.4(8).

8.16.3. Lines AB and CD, Fig. 8.4(9).

8.16.4. Lines AB and CD, Fig. 8.4(10).

8.16.5. Lines AB and CD, Fig. 8.4(11).

8.16.6. Lines AB and CD, Fig. 8.4(12).

Group 17. Shortest distance between two lines. Plane method

In each problem find the location, true length, and true slope of the shortest distance between the two given lines. Use the plane method of Section 3.13.

8.17.1. Lines AB and CD, Fig. 8.4(7).

8.17.2. Lines AB and CD, Fig. 8.4(8).

8.17.3. Lines AB and CD, Fig. 8.4(9).

8.17.4. Lines AB and CD, Fig. 8.4(10).

8.17.5. Lines AB and CD, Fig. 8.4(11).

8.17.6. Lines AB and CD, Fig. 8.4(12).

Group 18. Shortest level line between two lines

In each problem find the location, true length, and bearing of the shortest level line to connect the two given lines.

8.18.1. Lines AB and CD, Fig. 8.4(7).

8.18.2. Lines AB and CD, Fig. 8.4(8).

8.18.3. Lines AB and CD, Fig. 8.4(9).

8.18.4. Lines AB and CD, Fig. 8.4(10).

8.18.5. Lines AB and CD, Fig. 8.4(11).

8.18.6. Lines AB and CD, Fig. 8.4(12).

Group 19. Shortest line of given slope between two lines

In each problem find the location, true length, and bearing of the shortest line of given slope to connect the two given lines.

8.19.1. Lines AB and CD, 30°, Fig. 8.4(7).

8.19.2. Lines AB and CD, 20°, Fig. 8.4(8).

8.19.3. Lines AB and CD, 10°, Fig. 8.4(9).

Group 20. Line piercing a plane

In each problem show in both the given views where the given line pierces the given plane. Solve using two views only, and check by the edge-view method.

8.20.1. Line MN, plane ABCD, Fig. 8.4(2).

8.20.2. Line MN, plane ABC, Fig. 8.4(3).

8.20.3. Line MN, plane ABC, Fig. 8.4(4).

8.20.4. Line MN, plane ABC, Fig. 8.4(5).

8.20.5.   Line *HK*, plane *ABC*, Fig. 8.4(5).

8.20.6.   Line *MN*, plane *ABC*, Fig. 8.4(6).

Group 21. Intersection of two planes

In each problem find the line of intersection of the two specified planes, which should be placed beside each other and close together. Locate points on the line of intersection by all three methods.

I. Piercing-point method.

II. Auxiliary vertical-cutting-plane method.

III. Auxiliary horizontal-cutting-plane method.

8.21.1.   Plane *ABC* in Fig. 8.3(11) with plane *ABC* in Fig. 8.3(12).

8.21.2.   Plane *ABC* in Fig. 8.4(1) with plane *ABCD* in Fig. 8.4(2).

8.21.3.   Plane *ABC* in Fig. 8.4(1) with plane *ABC* in Fig. 8.4(3).

8.21.4.   Plane *ABC* in Fig. 8.4(1) with plane *ABC* in Fig. 8.4(4).

8.21.5.   Plane *ABC* in Fig. 8.4(3) with plane *ABC* in Fig. 8.4(6).

8.21.6.   Plane *ABC* in Fig. 8.4(5) with plane *ABCD* in Fig. 8.4(2).

Group 22. Dihedral angle

In each problem draw the two planes as they are specified, by connecting the given points. Find the true size of the dihedral angle between the two specified planes.

8.22.1.   Planes *ABX* and *ABC*, Fig. 8.3(10).

8.22.2.   Planes *ABX* and *ABC*, Fig. 8.3(11).

8.22.3.   Planes *ABC* and *BCX*, Fig. 8.3(12).

8.22.4.   Planes *ACX* and *ABC*, Fig. 8.3(11).

8.22.5.   Planes *ABC* and *BCD*, Fig. 8.4(7).

8.22.6.   Planes *ABC* and *ACD*, Fig. 8.4(8).

8.22.7.   Planes *ABD* and *ADC*, Fig. 8.4(9).

8.22.8.   Planes *ABC* and *ACD*, Fig. 8.4(10).

8.22.9.   Planes *ABD* and *BDC*, Fig. 8.4(11).

8.22.10.   Planes *ABD* and *ADC*, Fig. 8.4(12).

Group 23. Distance from a point to a plane

In each problem, using the two given views only, show the perpendicular from the given point to the given plane and show where this perpendicular pierces the plane. Check by drawing an edge view of the plane and measure the true length in this view.

8.23.1.   Point *X* to plane *ABC*, Fig. 8.3(10).

8.23.2.   Point *X* to plane *ABC*, Fig. 8.3(11).

8.23.3.   Point *X* to plane *ABC*, Fig. 8.3(12).

8.23.4.   Point *M* to plane *ABCD*, Fig. 8.4(2).

8.23.5.   Point *M* to plane *ABC*, Fig. 8.4(3).

8.23.6.   Point *M* to plane *ABC*, Fig. 8.4(4).

8.23.7.   Point *H* to plane *ABC*, Fig. 8.4(5).

8.23.8.   Point *N* to plane *ABC*, Fig. 8.4(5).

8.23.9.   Point *M* to plane *ABC*, Fig. 8.4(6).

8.23.10.   Point *C* to plane *ABD*, Fig. 8.4(8).

Group 24. Plane through a point perpendicular to a line

In each problem, using the two given views only, show a plane through the given point perpendicular to the given line.

8.24.1.   Point *X* and line *MN*, Fig. 8.3(2).

8.24.2.   Point *X* and line *MN*, Fig. 8.3(4).

8.24.3.   Point *X* and line *MN*, Fig. 8.3(8).

8.24.4.   Point *X* and line *MN*, Fig. 8.3(9).

Group 25. Project a line on to a plane

In each problem show, in both the given views, the projection of the specified line on to the specified plane.

8.25.1.   Line *MN* on plane *ABCD*, Fig. 8.4(2).

8.25.2.   Line *MN* on plane *ABC*, Fig. 8.4(3).

8.25.3.   Line *MN* on plane *ABC*, Fig. 8.4(4).

8.25.4.   Line *MN* on plane *ABC*, Fig. 8.4(5).

8.25.5.   Line *HK* on plane *ABC*, Fig. 8.4(5).

8.25.6.   Line *MN* on plane *ABC*, Fig. 8.4(6).

Group 26. Angle between a line and a plane

In each problem find the true size of the angle between the specified line and the specified plane.

8.26.1.   Line *MN* and plane *ABC*, Fig. 8.4(1).

8.26.2.   Line *MN* and plane *ABCD*, Fig. 8.4(2).

8.26.3.   Line *MN* and plane *ABC*, Fig. 8.4(3).

8.26.4.   Line *MN* and plane *ABC*, Fig. 8.4(4).

8.26.5.   Line *MN* and plane *ABC*, Fig. 8.4(5).

8.26.6.   Line *HK* and plane *ABC*, Fig. 8.4(5).

8.26.7.   Line *MN* and plane *ABC*, Fig. 8.4(6).

8.26.8.   Line *CD* and plane *ABC*, Fig. 8.4(7).

8.26.9.   Line *CD* and plane *ABC*, Fig. 8.4(8).

8.26.10.   Line *AB* and plane *ACD*, Fig. 8.4(9).

8.26.11.   Line *CD* and plane *ABC*, Fig. 8.4(10).

8.26.12.   Line *CD* and plane *ABC*, Fig. 8.4(11).

8.26.13.   Line *AB* and plane *ACD*, Fig. 8.4(12).

Group 27. To draw a plane figure on any oblique plane

In each problem show in the given views the specified figure lying in the specified plane. Assume the figure to be about half the size of the plane.

8.27.1.   In plane *ABC*, Fig. 8.3(10), show a square with two sides level.

8.27.2.   In plane *ABC*, Fig. 8.3(12), show an equilateral triangle with one side level.

8.27.3.   In plane *ABCD*, Fig. 8.4(2), show a hexagon having two sides parallel to the two given lines on the plane.

8.27.4.   In plane *ABC*, Fig. 8.4(4), show a square having two sides that are frontal lines.

8.27.5.   In plane *ABC*, Fig. 8.4(6), show an equilateral triangle with one side on the line *BC*.

8.27.6.   In plane *ABC*, Fig. 8.4(3), show a hexagon having one side parallel to the line *AB*.

Group 28. To draw a circle on any oblique plane

In each problem show in the given views a circle lying in the given plane with its center located as specified.

8.28.1.   Plane *ABC*, Fig. 8.4(3), center at *A*.

8.28.2.   Plane *ABC*, Fig. 8.4(5), inscribe the circle in the given triangular plane.

8.28.3.   Plane *ABC*, Fig. 8.3(11), center at *A*.

8.28.4.   Plane *ABCD*, Fig. 8.4(2), circle tangent to the two given lines, *AB* and *CD*.

Group 29. To show a solid object resting on an oblique plane

In each problem show the specified object on the plane in both the given views. "Standing" on the plane means that the axis is at right angles to the plane. "Lying" on the plane means that one face, not the base, coincides with the plane. The term "units" is used just to give some proportion between the altitude and the size of the base.

8.29.1   A right square prism standing on the plane ABC, Fig. 8.3(11). Two base edges are frontal lines. Altitude 1 unit, base 2 units.

8.29.2.   A right triangular prism standing on the plane $ABCD$, Fig. 8.4(2). One edge of the base is level. Base edge 1 unit, altitude 2 units.

8.29.3.   A right square prism lying on the plane ABC, Fig. 8.4(3). Axis of prism parallel to AB. Base edge 1 unit, altitude 2 units.

8.29.4.   A right square pyramid lying on the plane ABC, Fig. 8.4(4). One base edge is on the line BC. Base edge 1 unit, altitude 3 units.

8.29.5.   A right triangular pyramid standing on the plane ABC, Fig. 8.4(3). One base edge is level. Base edge 1 unit, altitude 2 units.

8.29.6.   A right triangular pyramid lying on the plane ABC, Fig. 8.4(5). One base edge is a frontal line. Base edge 1 unit, altitude 2 units.

8.29.7.   A thin circular disk standing on the plane ABC, Fig. 8.4(6).

8.29.8.   A thin circular disk lying on the plane ABC, Fig. 8.4(5). Axis parallel to the line AB.

Group 30. Geologist's compressed method

In each problem find the required answer using the geologist's compressed method. Given the strike and true dip of a plane, find the apparent dip in the specified direction.

8.30.1.   Strike N30°E, true dip 45°SE, find due east apparent dip.

8.30.2.   Strike S45°W, true dip 60°NW, find S60°W apparent dip.

8.30.3.   Strike N62°W, true dip 37°SW, find S73°W apparent dip.

In each of the following problems, given the strike and apparent dip of a plane, find the true dip.

8.30.4.   Strike N60°E, apparent dip 30° in due east direction.

8.30.5.   Strike S45°E, apparent dip 60° in N60°E direction.

8.30.6.   Strike N23°W, apparent dip 47° in S34°W direction.

In each of the following problems, given the strike and true dip of planes A and B, find the bearing and slope of their line of intersection.

8.30.7.   Plane A, N45°E, 30°SE; plane B, S60°E, 45°SW.

8.30.8.   Plane A, N30°W, 60°SW; plane B, N30°E, 45°SE.

8.30.9.   Plane A, S52°E, 28°SW; plane B, S15°W, 56°NW.

Group 31. Revolution of a point

In each problem revolve the given point about the given line as an axis and through the specified angle. Show the new position of the point in both the given views and call the new position X'. If there are two possible solutions, show only one of them.

8.31.1.   Fig. 8.3(1). Revolve X about MN through a 45° angle.

8.31.2.   Fig. 8.3(2). Revolve X about MN through a 90° angle.

8.31.3.   Fig. 8.3(3). Revolve X about MN through a 45° angle.

8.31.4.   Fig. 8.3(4). Revolve X about MN through a 30° angle.

8.31.5.   Fig. 8.3(5). Revolve X about MN through a 60° angle.

8.31.6.   Fig. 8.3(6). Revolve X about MN through a 90° angle.

8.31.7.   Fig. 8.3(7). Revolve X about MN through a 45° angle.

8.31.8.   Fig. 8.3(8). Revolve X about MN until it lies in the same vertical plane as MN.

8.31.9.   Fig. 8.3(9). Revolve X about MN until it appears on $m_2 n_2$.

8.31.10.   Fig. 8.3(5). Revolve X about MN until it lies at the same level as M.

Group 32. Revolution of a line

In each problem revolve the given line until it shows in its true length in the plan. Check the result by revolving the line until it shows in its true length in the front elevation.

8.32.1.   Line MN, Fig. 8.3(4).
8.32.2.   Line MN, Fig. 8.3(5).
8.32.3.   Line MN, Fig. 8.3(6).
8.32.4.   Line MN, Fig. 8.3(7).
8.32.5.   Line MN, Fig. 8.3(8).
8.32.6.   Line MN, Fig. 8.3(9).

Group 33. Revolution of a plane

In each problem revolve the given plane until it shows in its true size in the plan. Also revolve the plane until it shows in its true size in the front elevation, and see whether the two results check.

8.33.1.   Plane ABC, Fig. 8.3(10).
8.33.2.   Plane ABC, Fig. 8.3(11).
8.33.3.   Plane ABC, Fig. 8.3(12).
8.33.4.   Plane ABC, Fig. 8.4(1).
8.33.5.   Plane $ABCD$, Fig. 8.4(2).
8.33.6.   Plane ABC, Fig. 8.4(3).
8.33.7.   Plane ABC, Fig. 8.3(4).
8.33.8.   Plane ABC, Fig. 8.3(5).
8.33.9.   Plane ABC, Fig. 8.3(6).

In each of the following problems revolve the given plane about the specified axis until it shows in its true size in some view.

8.33.10.   Plane ABC, axis AB, Fig. 8.3(10).
8.33.11.   Plane ABC, axis BC, Fig. 8.3(11).
8.33.12.   Plane ABC, axis AB, Fig. 8.3(12).
8.33.13.   Plane ABC, axis BC, Fig. 8.4(1).
8.33.14.   Plane $ABCD$, axis CD, Fig. 8.4(2).
8.33.15.   Plane ABC, axis AB, Fig. 8.4(4).
8.33.16.   Plane ABC, axis BC, Fig. 8.4(6).

Group 34. Dihedral angle by revolution

In each problem find the true size of the dihedral angle between the two given planes by the method of revolution.

8.34.1.   Planes ABC and ABX, Fig. 8.3(10).
8.34.2.   Planes ABC and ABX, Fig. 8.3(11).
8.34.3.   Planes ABC and ABX, Fig. 8.3(12).
8.34.4.   Planes ABC and ABM, Fig. 8.4(3).
8.34.5.   Planes ABC and BCN, Fig. 8.4(6).
8.34.6.   Planes ABC and BCM, Fig. 8.4(5).
8.34.7.   Planes ABC and BCD, Fig. 8.4(7).
8.34.8.   Planes ABC and ABD, Fig. 8.4(9).

Group 35. Angle between a line and a plane by revolution

8.35.1.   Line MN and plane ABC, Fig. 8.4(1).
8.35.2.   Line MN and plane $ABCD$, Fig. 8.4(2).
8.35.3.   Line MN and plane ABC, Fig. 8.4(3).
8.35.4.   Line MN and plane ABC, Fig. 8.4(4).
8.35.5.   Line MN and plane ABC, Fig. 8.4(5).
8.35.6.   Line HK and plane ABC, Fig. 8.4(5).
8.35.7.   Line MN and plane ABC, Fig. 8.4(6).

Group 36. Cone locus

In each problem find the plan and elevation views of the line or lines satisfying the given angular conditions.

8.36.1.   A line having a 45° downward slope and making 60° with a level due east line.
8.36.2.   A line making 60° with a level line bearing due north and 30° with a level line bearing N45°E.
8.36.3.   A line from B making an angle of 45° with BA and 60° with BC, Fig. 8.3(10).

Group 37. Coplanar addition

In each problem find the magnitude and direction of the sum of the following concurrent coplanar vector quantities.

8.37.1. Two velocities: N60°E, 8 ft per second; S45°E, 6 ft per second. Show resultant.

8.37.2. Three forces: due north, 20 lb; S60°W, 40 lb; S45°E, 25 lb. Show equilibrant.

8.37.3. Four forces: S20°E, 400 lb; S75°W, 300 lb; N40°E, 550 lb; due south, 150 lb. Show resultant.

Group 38. Noncoplanar addition

In each problem find the magnitude, bearing, and slope of the sum of the following vector quantities.

8.38.1. Two forces: due east, 45° downward slope, 50 lb; N60°E, level, 100 lb. Show resultant.

8.38.2. Three velocities: N45°W, level, 30 ft per second; S30°W, 30° downward slope, 40 ft per second; due east, 60° downward slope, 20 ft per second. Show equilibrant.

8.38.3. Four forces: N50°E, 45° upward slope, 300 lb; S60°E, 20° downward slope, 400 lb; due south, level, 200 lb; due east, 40° downward slope, 250 lb. Show equilibrant.

Group 39. Noncoplanar forces

In each problem the only known force is given some value. Find the value of the unknown force in each member of the structure caused by the given load.

8.39.1. Three members of a tripod structure supporting a vertical load of 100 lb. Fig. 8.5(1).

8.39.2. Three members of a tripod structure supporting a vertical load of 50 lb. Fig. 8.5(2).

8.39.3. Three members of a structural frame resisting a horizontal thrust load of 400 lb. Fig. 8.5(3).

8.39.4. Three members of a structural frame resisting a horizontal thrust load of 200 lb. Fig. 8.5(4).

8.39.5. Three members of a structural frame resisting two horizontal thrust loads of 100 lb each. Fig. 8.5(5).

8.39.6. Three members of a structural frame resisting a level pull of 400 lb and a vertical force of 500 lb. Fig. 8.5(6).

8.39.7. Three members of a structural frame supporting a hanging load of 600 lb. Fig. 8.6(1).

8.39.8. Three members of a structural frame resisting an angular thrust load of 100 lb. Fig. 8.6(2).

8.39.9. Three members in compression; the force on each should be given some value. Find the direction and the value of the load which would cause the given forces in the members. Fig. 8.6(3).

Group 40. Helix

8.40.1. Show two views of a right-hand cylindrical helix with a level axis. Show only one turn about the cylinder, and use 16 points to determine the curve.

8.40.2. Show two views of a left-hand cylindrical helix with a vertical axis. Show only one turn about the cylinder, and use 16 points to determine the curve.

8.40.3. Show two views of a right-hand conical helix. The cone of revolution has a vertical axis. Assume the lead to be one-third of the altitude of the cone. Show the three turns of the helix.

8.40.4. Show two views of a left-hand conical helix. The cone of revolution has a level axis. Assume the lead to be half the altitude of the cone. Show the entire helix.

Group 41. Cylinder representation

In each problem show the plan and front elevation of a right cylinder of revolution having a specified line for an axis. The relative proportions of the cylinder are given in each problem.

8.41.1. Diameter 2 units, altitude 1 unit. Axis line *MN*, Fig. 8.3(2).

8.41.2. Diameter 1 unit, altitude 2 units. Axis line *MN*, Fig. 8.3(4).

8.41.3. Diameter 1 unit, altitude 1 unit. Axis line *MN*, Fig. 8.3(5).

8.41.4. Diameter 1 unit, altitude 2 units. Axis line *MN*, Fig. 8.3(7).

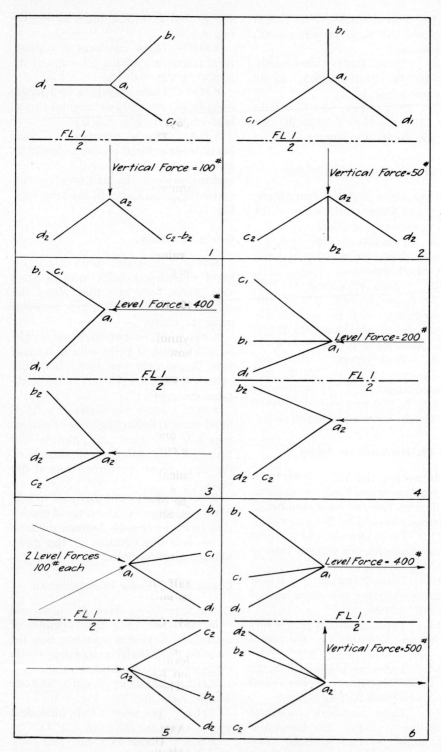

FIG. 8.5

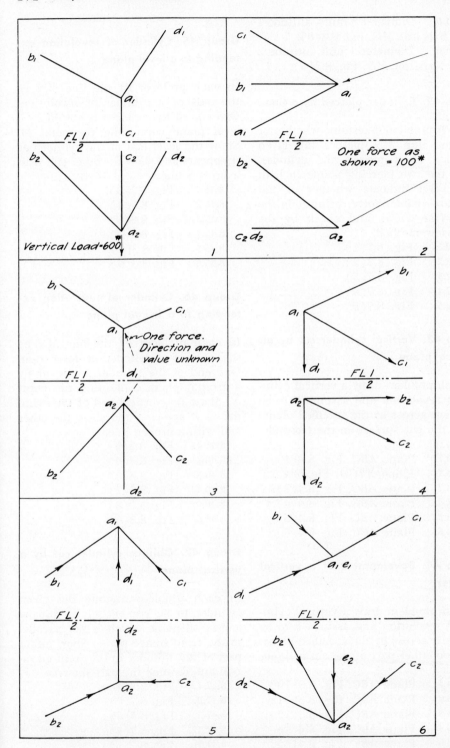

FIG. 8.6

8.41.5. Diameter 3 units, altitude 1 unit. Axis line *MN*, Fig. 8.3(8).

8.41.6. Diameter 1 unit, altitude 2 units. Axis line *MN*, Fig. 8.3(9).

Group 42. Cylinder pierced by a line

In each problem determine whether or not the line *MN* pierces the given cylinder. If it pierces the cylinder, show the two piercing points in both views and indicate whether or not they are visible. Solve, using only the two given views and check by the edge-view method.

8.42.1.	Fig. 8.7(1).
8.42.2.	Fig. 8.7(2).
8.42.3.	Fig. 8.7(3).
8.42.4.	Fig. 8.7(4).
8.42.5.	Fig. 8.7(5).

Group 43. Vertical cylinder cut by an oblique plane

In each problem draw a vertical cylinder of revolution and assume it to be cut clear across by the specified plane. Show the cut surface in the front elevation.

8.43.1.	Plane *ABC*, Fig. 8.4(1).
8.43.2.	Plane *ABCD*, Fig. 8.4(2).
8.43.3.	Plane *ABC*, Fig. 8.4(3).
8.43.4.	Plane *ABC*, Fig. 8.4(4).
8.43.5.	Plane *ABC*, Fig. 8.4(5).
8.43.6.	Plane *ABC*, Fig. 8.4(6).

Group 44. Development of a vertical cylinder

In each problem draw a vertical cylinder of revolution and assume it to be cut clear across by the specified plane. Develop either part of the cut cylinder surface.

8.44.1.	Plane *ABC*, Fig. 8.3(10).
8.44.2.	Plane *ABC*, Fig. 8.3(11).
8.44.3.	Plane *ABC*, Fig. 8.3(12).
8.44.4.	Plane *ABC*, Fig. 8.4(1).
8.44.5.	Plane *ABC*, Fig. 8.4(5).
8.44.6.	Plane *ABC*, Fig. 8.4(6).

Group 45. Cylinder of revolution extending to a level plane

In each problem let the line *MN* be the axis of a cylinder of revolution. One end of the cylinder is cut off by a level plane containing the point *M*. Show the level end of the cylinder as it appears and show the other end with a broken line.

8.45.1.	Fig. 8.3(3).
8.45.2.	Fig. 8.3(4).
8.45.3.	Fig. 8.3(5).
8.45.4.	Fig. 8.3(7).
8.45.5.	Fig. 8.3(8).
8.45.6.	Fig. 8.3(9).

Group 46. Cylinder of revolution extending to a frontal plane

In each problem let the line *MN* be the axis of a cylinder of revolution. One end of the cylinder is cut off by a frontal plane containing the point *N*. Show the vertical end of the cylinder as it appears, and show the other end with a broken line.

8.46.1.	Fig. 8.3(2).
8.46.2.	Fig. 8.3(4).
8.46.3.	Fig. 8.3(5).
8.46.4.	Fig. 8.3(7).
8.46.5.	Fig. 8.3(8).
8.46.6.	Fig. 8.3(9).

Group 47. Oblique cylinder cut by a vertical plane

In each problem assume the given cylinder to be cut entirely across by a vertical plane which is not parallel to the front image plane. Show either part of the cylinder in the front elevation and develop the part shown.

8.47.1.	Fig. 8.7(1).
8.47.2.	Fig. 8.7(2).
8.47.3.	Fig. 8.7(3).
8.47.4.	Fig. 8.7(4).
8.47.5.	Fig. 8.7(5).
8.47.6.	Fig. 8.7(6).

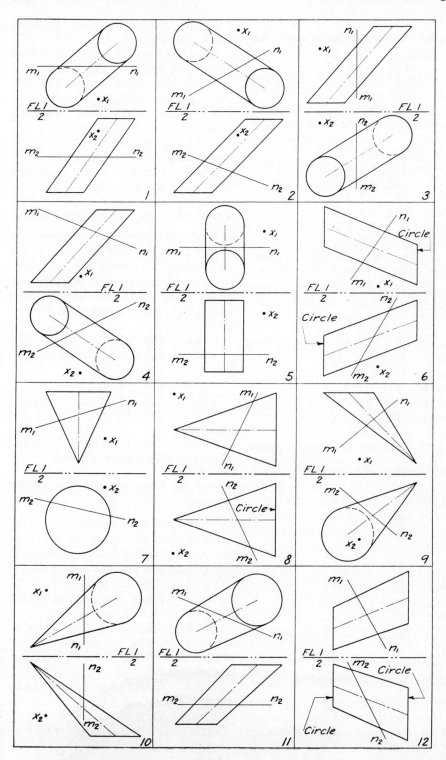

FIG. 8.7

Group 48. Oblique cylinder cut by an oblique plane

In each problem assume the given cylinder to be cut clear across by a sloping plane appearing as an edge in the front elevation. Show either part of the cylinder in the plan and develop the part shown.

8.48.1.   Fig. 8.7(1).
8.48.2.   Fig. 8.7(2).
8.48.3.   Fig. 8.7(3).
8.48.4.   Fig. 8.7(4).
8.48.5.   Fig. 8.7(5).
8.48.6.   Fig. 8.7(6).

In each problem assume the given cylinder to be cut clear across by the plane MNX which is unlimited in extent. Show either part of the cylinder in both the given views, and develop the part shown.

8.48.7.   Fig. 8.7(1).
8.48.8.   Fig. 8.7(2).
8.48.9.   Fig. 8.7(3).
8.48.10.   Fig. 8.7(4).
8.48.11.   Fig. 8.7(5).
8.48.12.   Fig. 8.7(6).

Group 49. Cylinder and tangent plane

In each problem show in the two given views a plane which is tangent to the given cylinder and contains the point X. Solve, using only two views, and check by a view showing the tangent plane as an edge.

8.49.1.   Fig. 8.7(1).
8.49.2.   Fig. 8.7(2).
8.49.3.   Fig. 8.7(3).
8.49.4.   Fig. 8.7(4).
8.49.5.   Fig. 8.7(5).
8.49.6.   Fig. 8.7(6). Extra view allowed.

Group 50. Cone representation

In each problem show the two views of a right cone of revolution having a specified line as an axis. The relative proportions of the cone are given for each problem. Place the vertex of the cone at the point M.

8.50.1.   Diameter 1 unit, altitude 3 units. Axis line MN, Fig. 8.3(3).
8.50.2.   Diameter 2 units, altitude 2 units. Axis line MN, Fig. 8.3(4).
8.50.3.   Diameter 3 units, altitude 2 units. Axis line MN, Fig. 8.3(6).
8.50.4.   Diameter 1 unit, altitude 2 units. Axis line MN, Fig. 8.3(7).
8.50.5.   Diameter 1 unit, altitude 3 units. Axis line MN, Fig. 8.3(8).
8.50.6.   Diameter 1 unit, altitude 2 units. Axis line MN, Fig. 8.3(9).

Group 51. Cone pierced by a line

In each problem determine whether or not the given line MN pierces the given cone. If it pierces the cone, show the two piercing points in both the given views, indicating whether they are visible. Solve, using only the two given views.

8.51.1.   Fig. 8.7(7).
8.51.2.   Fig. 8.7(8). Extra view allowed.
8.51.3.   Fig. 8.7(9).
8.51.4.   Fig. 8.7(10).
8.51.5.   Fig. 8.7(11). Vertex not available.
8.51.6.   Fig. 8.7(12). Vertex not available. Extra view allowed.

Group 52. Cone of revolution cut by a plane

In each problem assume the given cone of revolution to be cut across by the given plane. Show the cut in both the given views and find its true shape.

8.52.1.   Cone in Fig. 8.7(7), cut by a vertical plane not parallel to the base or containing the vertex.
8.52.2.   Cone in Fig. 8.7(7), cut by a plane showing as an edge in the front elevation and not containing the vertex.
8.52.3.   Cone in Fig. 8.7(7), cut by the plane MNX.
8.52.4.   Cone in Fig. 8.7(8), cut by a vertical plane not parallel to the base or containing the vertex.
8.52.5.   Cone in Fig. 8.7(8), cut by a sloping plane showing as an edge in

the front elevation and not containing the vertex.

8.52.6.   Cone in Fig. 8.7(8), cut by the plane *MNX*.

Group 53. Development of a cone of revolution

In each problem draw a right cone of revolution with the axis vertical. Assume it to be cut by the specified plane. Show the cut in both the given views and develop the lower or base end of the cone.

8.53.1.   Plane *ABC*, Fig. 8.4(1).
8.53.2.   Plane *ABCD*, Fig. 8.4(2).
8.53.3.   Plane *ABC*, Fig. 8.4(3).
8.53.4.   Plane *ABC*, Fig. 8.4(4).
8.53.5.   Plane *ABC*, Fig. 8.4(5).
8.53.6.   Plane *ABC*, Fig. 8.4(6).

Group 54. Development of an oblique cone with the vertex available

8.54.1.   Develop the entire cone of Fig. 8.7(9).

8.54.2.   Develop the entire cone of Fig. 8.7(10).

8.54.3.   Develop the lower portion of the cone of Fig. 8.7(9) after it is cut by a vertical plane not parallel to the base or containing the vertex.

8.54.4.   Develop the lower portion of the cone of Fig. 8.7(10) after it is cut by a vertical plane not containing the vertex.

8.54.5.   Develop the upper portion of the cone of Fig. 8.7(9) after it is cut by a sloping plane showing as an edge in the front view and not containing the vertex.

8.54.6.   Develop the upper portion of the cone in Fig. 8.7(10) after it is cut by a sloping plane showing as an edge in the front view and not containing the vertex.

8.54.7.   Develop the lower portion of the cone of Fig. 8.7(9) after it is cut by the plane *MNX*.

8.54.8.   Develop the upper portion of the cone of Fig. 8.7(10) after it is cut by the plane *MNX*.

Group 55. Development of an oblique cone with the vertex unavailable

8.55.1.   Develop the partial cone of Fig. 8.7(11).

8:55.2.   Develop the partial cone of Fig. 8.7(12).

Group 56. Cone and tangent plane

In each problem show in the two given views a plane which is tangent to the given cone and contains the point *X*. Solve, using only the two given views, and check by a view showing the tangent plane as an edge.

8.56.1.   Fig. 8.7(7).
8.56.2.   Fig. 8.7(8). Extra view allowed.
8.56.3.   Fig. 8.7(9).
8.56.4.   Fig. 8.7(10).

Group 57. Convolute

8.57.1.   Draw the plan and front elevation of a convolute generated by a line remaining tangent to a left-hand cylindrical helix. The cylinder axis is vertical. Assume the lead equal to half the diameter. Show one complete sweep of the convolute extending to a level plane. Develop one complete turn. Calculate the number of turns that may be developed in one piece.

8.57.2.   Draw the plan and front elevation of a convolute whose generatrix is tangent to a right-hand cylindrical helix. The axis of the cylinder is level. Assume the lead equal to the diameter. Show one complete turn extending to the plane of one of the bases. Develop one complete turn. Calculate the number of turns that may be developed in one piece.

8.57.3.   Use the same data as for problem 8.57.2. Let the convolute extend only to a larger cylinder twice as large in diameter as the first cylinder. Develop one complete turn of the portion of the convolute between the two cylinders.

Group 58. Helicoid

8.58.1.   Draw two views of the helicoidal surface of a right-hand square thread. Assume the outside diameter about 2 in. and the lead about ¾ in. Show the thread for two complete turns.

8.58.2.   Draw two views of the helicoidal surface of a triple right-hand square thread. Assume the outside diameter 2 in. and the lead 1½ in. Show the triple threads for only one turn.

Group 59. Hyperbolic paraboloid

In each problem draw three views of the hyperbolic paraboloid surface extending between the two given lines. The two lines may be extended at either end if it is necessary. The plane director will be specified in each case. Show about eight elements in each view.

8.59.1.   Lines *AB* and *CD*, Fig. 8.8(1). Plane director is the front image plane.

8.59.2.   Lines *AB* and *CD*, Fig. 8.8(2). Plane director is a level plane.

8.59.3.   Lines *AB* and *CD*, Fig. 8.8(3). Plane director is a profile plane.

8.59.4.   Lines *AB* and *CD*, Fig. 8.8(4). Plane director is a level plane.

8.59.5.   Lines *AB* and *CD*, Fig. 8.4(8). Plane director is the front image plane.

8.59.6.   Lines *AB* and *CD*, Fig. 8.4(10). Plane director is a level plane.

8.59.7.   Lines *AB* and *CD*, Fig. 8.4(11). Plane director is a profile plane.

8.59.8.   Lines *AB* and *CD*, Fig. 8.4(12). Plane director is a level plane.

Group 60. Conoid

8.60.1.   Draw three views of a conoid surface connecting the circle and the line *AB* in Fig. 8.8(5). The plane director is level. Show eight elements in each view.

8.60.2.   Draw three views of a conoid surface connecting the circle and the line *AB* in Fig. 8.8(6). The plane director is the front image plane. Show eight elements in each view.

Group 61. Cylindroid

8.61.1.   Draw three views of a cylindroid surface connecting the two curved lines in Fig. 8.8(7). The plane director is level. Show eight elements in each view.

8.61.2.   Draw three views of a cylindroid surface connecting the two curved lines in Fig. 8.8(8). The plane director is the front image plane. Show eight elements in each view.

Group 62. Hyperboloid of revolution of one sheet

In each problem draw the two simplest views of the hyperboloid of revolution of one sheet which would be generated by revolving the given line *CD* about the given line *AB* as an axis. Show twelve elements in each view.

8.62.1.   Fig. 8.8(1).
8.62.2.   Fig. 8.8(2).
8.62.3.   Fig. 8.8(3).
8.62.4.   Fig. 8.8(4).
8.62.5.   Fig. 8.4(7).

Group 63. Sphere

In each problem find in both the given views the two points at which the specified line pierces the given sphere.

8.63.1.   Line *AB*, Fig. 8.8(9).
8.63.2.   Line *CD*, Fig. 8.8(9).
8.63.3.   Line *MN*, Fig. 8.8(9).

In each problem show in both the given views the cut made by the specified plane on the spherical surface. Also show the true size of the cut.

8.63.4.   Plane No. 1, Fig. 8.8(10).
8.63.5.   Plane No. 2, Fig. 8.8(10).
8.63.6.   Plane *ABC*, Fig. 8.8(10).

In each problem show in the two

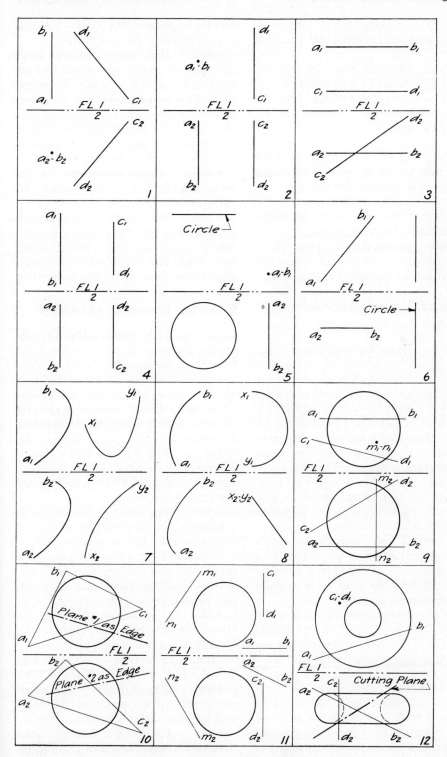

FIG. 8.8

given views a plane that is tangent to the sphere and contains the specified line.

8.63.7.   Line *AB*, Fig. 8.8(11).
8.63.8.   Line *CD*, Fig. 8.8(11).
8.63.9.   Line *MN*, Fig. 8.8(11).

Group 64. Torus

8.64.1.   In Fig. 8.8(12) show in the two given views the points at which the line *CD* pierces the torus.

8.64.2.   In Fig. 8.8(12) show in the two given views the points at which the line *AB* pierces the torus.

8.64.3.   In Fig. 8.8(12) show in the plan view the sectional cut across the torus made by the given cutting plane.

Group 65. Double-curved surfaces of revolution

8.65.1.   Draw three views of a paraboloid of revolution which is generated by revolving a parabola whose equation is $y^2 = 4x$. The axis of revolution is a symmetrical axis containing the focus. Calculate values of y for values of x from 1 to 6. Plot the curve to some scale. Assume the base to be a right section where x equals 6.

8.65.2.   Draw three views of a paraboloid of revolution which is generated by revolving a parabola whose equation is $x^2 = 8y$. Take the axis to be a vertical line through the focus. Calculate values of x for values of y from 1 to 6. Plot the curve to some scale. Assume the base to be a right section where y equals 6.

8.65.3.   Draw three views of a hyperboloid of revolution of two sheets which is generated by revolving a hyperbola whose equation is $(x^2/9) - (y^2/16) = 1$. Use a level axis through the foci.

Calculate values of y for values of x from 3 to 8. Plot the curve to some scale and make the bases right sections where x equals +8 or −8.

8.65.4.   Same statement as for problem 8.65.3, but let the equation for the hyperbola be $(y^2/9) - (x^2/16) = 1$. Take the axis vertical.

8.65.5.   Draw the three views of an ellipsoid of revolution generated by revolving an elipse about its major axis. The equation of the ellipse is $(x^2/16) + (y^2/9) = 1$. Calculate the lengths of the major and minor axes and draw the ellipse by the trammel method. Take both axes to be level.

8.65.6.   Same statement as for Prob. 8.65.5, but let the ellipse revolve about its minor axis.

8.65.7.   An ellipsoid of revolution is generated by an ellipse having a major axis 3 units long and a minor axis 2 units long. The major axis is a vertical line. Draw three views of the surface.

Group 66. Intersection of surfaces

In each problem find the line of intersection between the two given surfaces located as shown in the drawing. Show the hidden as well as the visible part of the curve and locate all special points of tangency. Show each extreme element with a solid line as far as it is visible.

8.66.1.   Fig. 8.9(1). Two elliptical cylinders with axes nonintersecting. Also develop one of the cylinders and the line of intersection.

8.66.2.   Fig. 8.9(2). Two elliptical cylinders. Also develop one of the cylinders and the line of intersection.

8.66.3.   Fig. 8.9(3). Cylinder of revolution and oblique pyramid. Also develop both surfaces and show the line of intersection on each one.

8.66.4.   Fig. 8.9(4). Two oblique cones. Show two possible methods for obtaining points on the curve of intersection. Also develop one of the cones and the line of intersection.

8.66.5.   Fig. 8.9(5). Two oblique cones. Also develop one of the cones and the line of intersection.

8.66.6.   Fig. 8.9(6). Cone of revolution and triangular pyramid. Also develop both surfaces and show the line of intersection on each one.

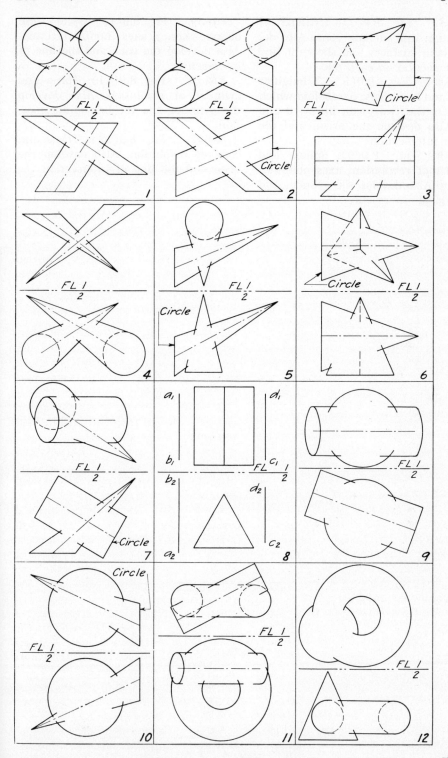

FIG. 8.9

8.66.7.   Fig. 8.9(7). Cylinder of revolution and oblique cone. Also develop both surfaces and show the line of intersection on each one.

8.66.8.   Fig. 8.9(8). Triangular prism and hyperbolic paraboloid which connects the two given lines *AB* and *CD*. Also develop the prism and the line of intersection.

8.66.9.   Fig. 8.9(9). Sphere and cylinder of revolution, axes nonintersecting. Also develop the cylinder and the line of intersection.

8.66.10.   Fig. 8.9(10). Sphere and oblique cone, axes nonintersecting. Also develop the cone and the line of intersection.

8.66.11.   Fig. 8.9(11). Torus and cylinder of revolution. Also develop the cylinder and the line of intersection.

8.66.12.   Fig. 8.9(12). Torus and cone of revolution. Also develop the cone and the line of intersection.

9

APPLIED PROBLEMS[1]

The problems in this chapter are all given with definite working data including the proper scale to use for each one. Practically all of them have engineering settings and a large number of them have been suggested by engineers who have had to solve similar problems in their own work. Every problem should be solved accurately for a result which should be scaled for a numerical value wherever possible. Required angles should be measured with a protractor to the nearest half of a degree, or the tangent of the angle should be measured. In all cases the results should be plainly marked.

The following fourfold experience will be gained by a careful solution of these problems.

1. Experience in laying out problems from data which closely resemble the data in an engineering office.

2. Experience in becoming familiar with engineering terms, in analyzing engineering situations, and in selecting the proper principles to be used in arriving at a correct solution.

[1] See the Warner and Douglass Problem Book, which has about 75 problems partly laid out ready to be solved.

3. Experience in establishing the habit of checking every problem by an independent method.

4. Experience in learning to work accurately in order to obtain proper results, and in learning to show your work clearly so that the drawing may easily be understood by others.

The problems are all laid out so that they can be solved on one of two standard-size sheets, using the scales that are given. The small sheet trims to a size 8½ by 11 in., and the large sheet trims to a size 11 by 16¼ in. There is a ¾-in. border on the left side of the large sheet and on the upper side of the small sheet, and a ¼-in. border on the other three sides of each sheet. The right edge of the large sheet will fold over to the left border line, and the sheet will file in a standard notebook with the small sheet. Both sheets are to be placed with the long edge parallel with the T square unless the problem number is followed by a *T*, which means to turn the sheet so that the short edge is parallel with the T square. All problems are to be solved on small sheets unless the problem number is followed by an *L*, which means to use a large sheet. North is always at the top of the sheet unless it is otherwise stated. All figures in this chapter have the same number as the problem they accompany.

Group 1. True length and true slope of a line

9.1.1. Portion of concrete wing wall. Scale: ¼ in. = 1 ft 0 in.

1. Find the true length of the corner *AB* in an inclined view.

2. Find the true slope of the corner *AB*.

9.1.2. Two guy wires for a stack. Scale: ½ in. = 1 ft 0 in.

1. Find the true length of the wire to *A*.

2. Find the true length of the wire to *B*.

3. Find the true slope of the wire to *A*.

4. Find the true slope of the wire to *B*.

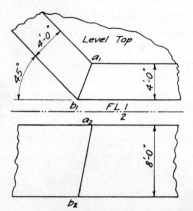

PROB. 9.1.1

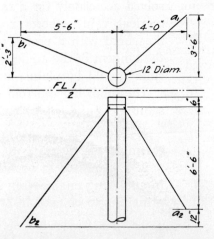

PROB. 9.1.2

5. Check both true lengths by new views.

9.1.3.   Scale: ¼ in. = 1 ft 0 in.

The line *AB* is the centerline of a metal chute the same length as *AB*. The end view of the chute is as shown, and in this view the centerline appears as a point and is located as shown. Both ends of the chute are cut off square, and it is to be placed so that the 3-ft faces are vertical planes.

Show the complete chute in the given views, including all hidden lines.

9.1.4.   Scale: 1 in. = 50 ft.

The points *C* and *A* are the portals of two tunnels. *C* is 200 ft east and 50 ft south of *A* and 50 ft lower in elevation than *A*.

From *A* one tunnel bears N60°E and falls 45° (true slope).

From *C* the other tunnel bears N45°W until it meets the tunnel from *A*.

Find:

1. The true slope of the tunnel from *C*.
2. The true length of the tunnel from *C*.
3. The true length of the tunnel from *A*.

9.1.5.   Scale: 1 in. = 100 ft.

A tunnel 300 ft long is driven from the point *A* bearing N60°E and on a falling grade of 100 per cent. At what point on the map would you start a second tunnel bearing N45°W and on a falling grade of 57.7 per cent so that it would meet the low end of the first tunnel? Assume that the two tunnels start at the same level.

Find:

1. The true length of the second tunnel.
2. The starting point of the second tunnel on the map with reference to *A*. Give coordinates.

9.1.6.   Scale: 1 in. = 50 ft.

Two tunnels start from a common point *A* in a vertical shaft. Tunnel *AB* bears N45°W, falls 15 per cent, and is 150 ft long. Tunnel *AC* bears S75°W, falls 15°, and is 178 ft long. The ends of the two tunnels are to be

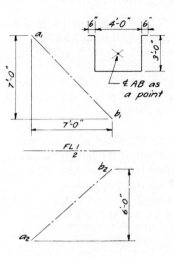

PROB. 9.1.3

connected by a ventilating tunnel.

Find the following information regarding the ventilating tunnel:

1. The bearing.
2. The true length.
3. The per cent grade.
4. The true slope in degrees.

9.1.7.   Scale: 1 in. = 100 ft. Given a pipeline located as follows:

Line	Course	Slope	True length, feet
AB	Due E	−100 per cent	200
BC	S45°E	+30°	150
CD	S30°W	−0.5	200

(See Figs. 2.7 and 2.8 for the three ways of indicating slope.)

1. Draw a plan view of the entire pipeline and auxiliary elevations of each portion of the line.
2. It is desired to replace these pipes with one straight pipe direct from *A to D*. Find the bearing, true length, and true slope (in degrees) of this new pipe, *AD*.

9.1.8.   A vertical mast, *DE*, and three guy wires to anchors *A*, *B*, and *C*. Scale: ⅛ in. = 1 ft 0 in.

Anchor *A* is 14 ft W and 4 ft N of *DE* and at an elevation of 140 ft.

Anchor *B* is 12 ft E and 7 ft N of *DE* and at an elevation of 145 ft.

Anchor C is 4 ft E and 10 ft S of DE and at an elevation of 138 ft.

Elevation of $D = 158$ ft and of $E = 143$ ft. All elevations are referred to sea level. The boom is pivoted at E and is to swing clear around the mast at any angle of elevation.

Find:

1. The true length of each guy wire.
2. The length of the longest boom it is possible to use and have it clear all the guy wires.
3. The length of the longest horizontal boom that can be used.

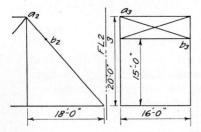

PROB. 9.1.9

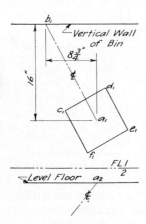

PROB. 9.1.10

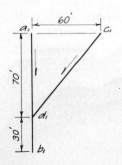

PROB. 9.1.11

4. Keeping the anchor C on the same level, how far would it have to be moved due south so that a boom 9 ft long could be used in any position?

9.1.9.   End panel of a bridge. Scale: ⅛ in. = 1 ft 0 in.

Two elevation views are given.

Find:

1. The true length of the diagonal member AB.
2. The true slope of the diagonal member AB in degrees.

9.1.10.   Scale: 1½ in. = 1 ft 0 in.

A hopper discharges into an 8-in. square opening in a level floor. AB is the centerline of a metal chute which is to carry material from the hopper to the vertical wall of a bin. AB has a true slope of 30°.

1. Find the true size of the hole in the vertical wall.
2. Find the true size and shape of a right section (or end view) of the chute.
3. What is the per cent reduction in area from the hopper bottom to the right sectional area of the chute?
4. How many inches would you change the elevation of B to comply with the specification that the reduction in area must be only 25 per cent? Show the end view of the chute in this new position.

9.1.11.   Scale: 1 in. = 40 ft.

Two pipes, AB and CD, intersect at D. They both slope downward in the direction of the arrows as shown in the plan view. Both pipes have the same slope, and the true length of pipe AB is 130 ft.

1. Find the difference in elevation between points A and C.
2. Find the true length of pipe CD.
3. Show both pipes in the front elevation.

9.1.12.   Scale: ¼ in. = 1 ft 0 in.

An enclosed four-sided metal chute is to be installed to connect the hole in the bottom of the bin with the top

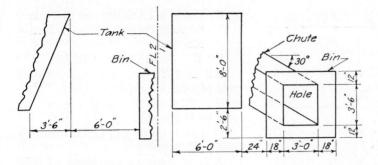

PROB. 9.1.12

of the tank. The chute is partially shown in the plan. The four long corners of the chute are parallel and have a true slope of 37°.

1. Find the true size and shape of the hole to be cut in the top of the tank.
2. Draw a view showing the true size and shape of a square cut across the chute. A square cut is at right angles to all four long edges.
3. Show the entire chute in every view.

9.1.13.   Scale: 1 in. = 40 ft.

The drawing shows the plan view of a structural-steel microwave reflector tower 120 ft tall having a cross section in the shape of an equilateral triangle 5 ft on a side. It is steadied by 12 tie downs anchored in level ground. The tower tapers to a point in the top 30 ft. Cables are attached to the nearest edge of the tower at points A, B, C, and D at elevations of 120, 90, 60, and 30 ft, respectively, above the ground. Show the front elevation of the tower and ties and find the true length and slope of typical ties AF, BF, CE, and DE.

9.1.14.[T]   Scale: ⅛ in. = 1 ft 0 in.

CD is the longitudinal centerline of a right circular cylindrical tank of length CD. Point P is on the surface of the cylinder.

D is 20 ft east and 11 ft south of C, and 7 ft lower than C.

P is 11 ft east and 12 ft south of C, and 2 ft lower than C.

Find the length and diameter of the tank and show the entire tank in all views. Plot the elliptical ends in the plan and elevation views by projecting 8 points in each ellipse.

9.1.15.   Scale: ¼ in. = 1 in.

Spur gears mesh correctly with each other when their "pitch diameters" are just in contact. The centerline AB of a shaft carrying a 3 in. PD gear is given. Only the plan view is given of centerline CD of a shaft carrying an 8 in. PD gear. CD is parallel and equal in length to AB.

B is 4 in. east and 7 in. north of of A, and 3 in. higher than A.

C is 5 in. due east of A.

Locate the front elevation of CD so that the gear that it carries will mesh with the gear on AB. Choose the position of CD that is lower than AB. How much lower is C than A?

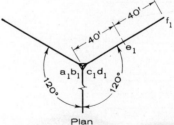

PROB. 9.1.13

Group 2. Edge view and true size of a plane

9.2.1.   Plane ABC. Scale: ¼ in. = 1 ft 0 in.

B is 4 ft east and 7 ft north of A and 5 ft below A.

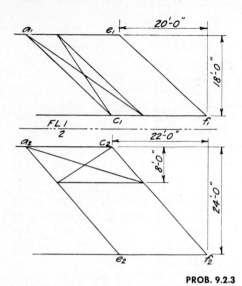

PROB. 9.2.3

C is 10 ft east and 4 ft north of A and 3 ft above A.

Find:

1. Edge view of the plane in an elevation view.
2. Edge view of the plane in an inclined view.
3. True slope of the plane.
4. True size of the plane off each edge view.
5. Show in all views a point X on this plane which is 9 ft away from C and at an elevation of 5 ft below C.

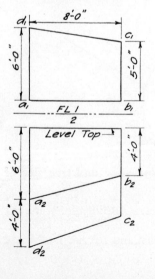

PROB. 9.2.5

6. Show a level line on the plane through the point X in each truesize view.

9.2.2.   Scale: 1 in. = 40 ft.

AB and BC are the centerlines of two pipelines intersecting at B.

A is 60 ft north and 60 ft west of B and 60 ft above B.

C is 30 ft north and 120 ft west of B and 20 ft below B.

X is a point on the plane of ABC 40 ft due west of A.

Find:

1. True size of plane angle ABC.
2. Where to connect into BC with a 45° connection for a straight pipe from X. Give the distance from B.
3. True length of the pipe from X to BC.
4. True slope of the pipe from X to BC.
5. Show the pipe from X to BC in all views.

9.2.3.   Skew bridge. Scale: $\frac{3}{32}$ in. = 1 ft 0 in.

Only the end panel, AEFC, of this bridge is given.

Draw a view of this entire panel in which it will appear in its true size and shape. The members AC and EF are parallel.

9.2.4.   Scale: ¼ in. = 1 ft 0 in.

It is desired to make the shortest possible connection from a point C to a pipe AB which bears due north from A. The point B is 10 ft due north of A and 8 ft higher than A. The point C is 6 ft due east of A and on the same level as B. Show the shortest connecting pipe in all views.

Find:

1. True length of this shortest pipe.
2. True slope of this shortest pipe.
3. True distance from A to the point of connection.
4. Check by an independent method.

9.2.5.   Scale: ¼ in. = 1 ft 0 in.

A metal tank to be placed below deck on a ship has a level top, four vertical sides and a sloping bottom

ABCD. AB and *CD* are not parallel but the bottom is a plane surface and *C* is on this plane. This fact may be used to locate *C* in the front elevation.

1. Draw an auxiliary elevation of the entire tank showing the bottom of the tank as an edge.
2. Draw an inclined view of the bottom of the tank only which will show it in its true size and shape.

9.2.6. Plane surface *ABC*. Scale: ¼ in. = 1 ft 0 in.

C is 9 ft south and 4 ft east of *A* and 8 ft below *A*.

B is 6 ft south and 10 ft east of *A* and 2 ft below *A*.

Show in the plan and front elevation the largest circle which may be laid out in the plane *ABC* and within the given limits of the plane.

Obtain the axes for each view direct from an edge view.

In a view showing the true size of the plane *ABC* show all four axes used for both views and label each one.

9.2.7.[L] Point of a concrete pier. Scale: ¼ in. = 1 ft 0 in.

The top and the bottom of the pier are level planes. Faces *A* and *B* have a true slope of 60°. Halfway up the face *A*, on the centerline shown, is the center of a recess 6 ft square and 18 in. deep. Two sides of the square opening are level. Halfway up the face *B*, on the centerline shown, is the center of a recess 6 ft in diameter and 18 in. deep. Both recesses are set in at right angles to the sloping faces.

Show both the recesses in both the given views.

9.2.8. Timber tower supporting tank. Scale for tower: ⅛ in. = 1 ft 0 in. Scale for angles: 1½ in. = 1 ft 0 in.

The four corner posts are 8 by 8 timber columns placed as shown in the sketch.

Find the angles of cut across the sides of the timber at the base which will make the end cut a level plane when the timber is in position.

9.2.9. A water main having nega-

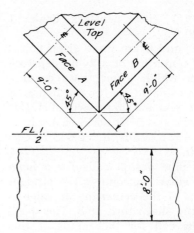

PROB. 9.2.7

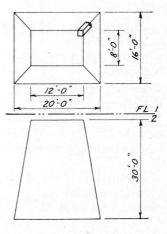

PROB. 9.2.8

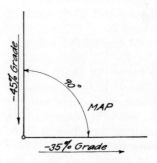

PROB. 9.2.9

tive grades is to turn the corner at a 90° street intersection as shown on the map.

Find the angle required for the special elbow at this corner.

9.2.10.[L] Steel penstock. Scale: 1 in. = 20 ft.

A, B, and C are three given points on the centerline of a 24-in. penstock through which water flows down a mountainside to the power house. B is 70 ft west and 25 ft north of A. C is 30 ft west and 50 ft north of A. Elevations: A = 2,675 ft, B = 2,655 ft, C = 2,630 ft.

The pipe design calls for a long-radius elbow to connect the straight portions of the centerline in order to avoid the sharp turn at B. The specifications call for the radius of this elbow to be eight times the diameter of the pipe.

Find:

1. True size of the sweep angle of the elbow.
2. True length of straight pipe from A to the elbow.
3. True slope of straight pipe from A to the elbow.
4. True length of straight pipe from the elbow to C.
5. True slope of straight pipe from the elbow to C.
6. Show the centerline of the elbow in the plan and front elevation.

9.2.11. Scale: 1 in. = 40 ft.

A, B, and C are three points on the centerline of a steel penstock delivering water down a mountainside.

B is 130 ft east and 20 ft south of A and 20 ft above A.

C is 50 ft east and 80 ft north of A and 70 ft above A.

In order to avoid the sharp turn at A, a special elbow is to be installed there. This elbow is to come tangent to AB and AC, where it connects with them. The radius of the elbow is 60 ft. Find the bearing from A to the center of curvature of the elbow. Show the elbow in all views.

9.2.12. Scale: 1 in. = 60 ft.

Problem 9.2.12 shows some parallel straight-line contour lines on the ground, and two definitely located points, A and B.

1. Lay out the centerline of a road

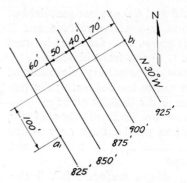

PROB. 9.2.12

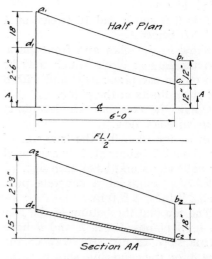

PROB. 9.2.13

from A to B so that it will have a constant grade of 10 per cent for the entire distance. Use no more switchbacks than are necessary and neglect the curves at the turns.

2. What would be the maximum grade if a road of constant bearing were used?

9.2.13. Flume transition. Scale: ½ in. = 1 ft 0 in.

The plane ABCD is one of the side walls of this transition which is to be made of reinforced concrete. Find the true size and shape of the form to be used for this side wall.

9.2.14. Scale: 1 in. = 5 ft.

Problem 9.2.14 shows the plan view

of two parallel pipes cut off at A and B. Pipe A is 6 ft higher than pipe B. Show how to connect the two pipe ends with a symmetrical ogee pipe. Determine the true radius for bending and the number of degrees of the bend angle.

Also show just the centerline of the ogee pipe in the plan.

9.2.15.   Scale: 1 in. = 5 ft.

Use the same pipes as in Prob. 9.2.14. The point A remains fixed, but the point B must be changed. Show how to connect these two pipes with a symmetrical ogee pipe having a bend radius of 12 ft.

1. Find how much will have to be cut off the pipe at B.
2. Find the number of degrees of the bend angle.
3. After solving graphically, check all results by a mathematical solution.

9.2.16.   Scale: ¼ in. = 1 ft 0 in.

A point source of radiation P and a rectangular surface are shown in Prob. 9.2.16. The surface is to be protected by a lead shield 2 ft away from and parallel to the surface. What is the size of the shield? Show the shield in all views.

9.2.17.   Scale: ½ in. = 1 in.

Problem 9.2.17 shows the plan view of the base of a metal bracket. The main bracket consists of the base and a second member, each ¾ in. thick and welded together. The front face of the second piece intersects the upper surface of the base at line AB and makes an angle of 60° with the horizontal, leaning away from the observer. Measuring along the slant face at right angles to AB, this piece is 4 in. long and is cut off square on the end. A vertical triangular supporting web ½ in. thick is placed behind and perpendicular to the slanting member at its mid-point with the vertex of the web at the top of the slanting member and its base run out to the extremity of the bracket base. Complete the plan and front elevation views of the bracket showing proper visibility.

9.2.18.   Scale: ¼ in. = 1 ft 0 in.
Plane surface $ABCD$.

B is 11 ft due east of A, and 4 ft above A.

C is 11 ft due east and 4 ft south of A, and 1 ft below A.

D is 5 ft east and 6 ft south of A, elevation not given.

Complete the front elevation view of $ABCD$ using only the two given

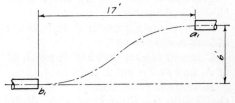

PROB. 9.2.14

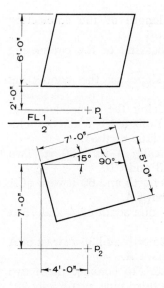

PROB. 9.2.16

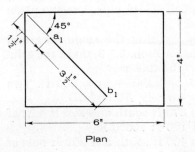

Plan

PROB. 9.2.17

views. Check the result by obtaining an edge view of *ABCD*.

1. Find the angle plane *ABCD* makes with the front elevation image plane.
2. Find the distance *D* is below *A*.

Group 3. Shortest distance between two lines

9.3.1.   Scale: 1 in. = 100 ft.

From *A* a new mining shaft bears N60°E and falls 30°.

From *C* an old mining shaft bears S75°W and falls 15°.

C is 140 ft east and 210 ft north of of *A*, and at the same level as *A*.

It is desired to connect these two shafts with the shortest possible shaft. Use the line method.

Find:

1. True length of the connecting shaft.
2. True bearing of the connecting shaft.
3. True slope of the connecting shaft.

9.3.2.   Using the same data and scale as for Prob. 9.3.1, make the solution by the plane method.

9.3.3.   Scale: 1 in. = 30 ft. Two pipelines, *AB* and *CD*.

B is 10 ft north and 60 ft west of *A* and 45 ft above *A*.

C is 22 ft due south of *A* and 75 ft above *A*.

D is 50 ft north and 35 ft west of *A* and 35 ft above *A*.

Show where to connect these two pipes with a third pipe, using only 90° tees. Find the true length, bearing, and slope of the third pipe. Use the plane method.

9.3.4.   Using the same data and scale as for Prob. 9.3.3, make the solution by the line method.

9.3.5.   Scale: 1 in. = 200 ft. Two mining tunnels, *AB* and *CD*.

B is 370 ft south and 220 ft east of *A* and 120 ft above *A*.

C is 400 ft south and 200 ft west of *A* and 490 ft above *A*.

D is 150 ft north and 120 ft east of *A* and 210 ft above *A*.

It is desired to connect these two tunnels with the shortest possible shaft. Show the shaft in all views.

Find:

1. True length of the new shaft.
2. True slope of the new shaft.
3. True bearing of the new shaft.

9.3.6.   Scale: 1 in. = 40 ft.

Two mining tunnels are driven from the points *A* and *C*. The point *C* is 50 ft east and 37 ft south of *A* and 60 ft lower than *A*. The tunnel from *A* bears due east and the one from *C* bears due north. Each tunnel is 100 ft long and has a −63 per cent grade.

Using the line method, find the true length, grade, and bearing of the shortest possible shaft to connect the two given tunnels.

9.3.7.   Scale: 1 in. = 400 ft. Two mining shafts, *AB* and *CD*.

B is 800 ft east and 600 ft south of *A* and 800 ft below *A*.

C is 100 ft east and 900 ft south of *A* and 600 ft below *A*.

D is 700 ft east and 100 ft south of *A* and 100 ft below *A*.

Connect these two shafts with the shortest possible level tunnel and show it in the plan and the front elevation.

Find:

1. True length of the level tunnel.
2. Bearing of the level tunnel.

9.3.8.   Using the same scale and layout data as for Prob. 9.3.1, draw the two mining shafts and connect them with the shortest possible level tunnel.

Find:

1. True length of the level tunnel.
2. Bearing of the level tunnel.

9.3.9.   Using the same scale and layout data as for Prob. 9.3.3, draw the two pipes and show how to connect them with the shortest possible level pipe.

Find:

1. True length of the level pipe.
2. Bearing of the level pipe.

9.3.10.   Using the same scale and

layout data as for Prob. 9.3.7, draw the two mining shafts and locate the shortest possible connecting shaft.

Find:

1. True length of the new shaft.
2. True slope of the new shaft.
3. Bearing of the new shaft.

9.3.11.   Use the same layout data and the same scale as for Prob. 9.3.3. Find the location and the true length of the shortest connecting pipe to bear N30°E and to have a rising 30 per cent grade.

9.3.12.   Scale: ⅛ in. = 1 ft 0 in.

A vertical mast 18 ft high and a wire AB are located as shown. A guy wire from the point X is to be fastened to the mast as high up as possible.

Find the highest point up the mast at which it could be fastened if the clearance between the guy wire and the wire AB must be at least 24 in.

9.3.13.   Scale: 1 in. = 40 ft.

Two mining tunnels are driven from points A and C. The point C is 50 ft east and 37 ft south of A and 60 ft lower than A. The tunnel from A bears due east, and the tunnel from C bears due north. Both tunnels have a -63 per cent grade. The tunnel from A is 118 ft long, and the tunnel from C is 94 ft long.

Locate the shortest possible tunnel to connect the two given tunnels and having a downward grade of 20 per cent from the A tunnel to the C tunnel.

1. Find the bearing and true length of this new tunnel.
2. Find the distance from A to the upper end of the new tunnel.
3. Find the distance from C to the lower end of the new tunnel.

9.3.14.   Use the same scale and data as for Prob. 9.3.3, except that for this problem AB and CD are two mining tunnels. In addition, the point M is the lower end of a vertical shaft which is 30 ft due west of A. This point M is at the same elevation as B. Locate a straight tunnel connecting the two tunnels AB and CD and passing through the point M.

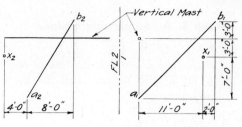

PROB. 9.3.12

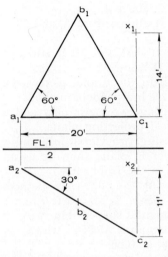

PROB. 9.3.15

Find the true length, bearing, and true slope of this new tunnel.

9.3.15.   Scale: ⅛ in. = 1 ft 0 in.

Problem 9.3.15 shows point X and triangular steel frame ABC. An electric power line runs from X in a S45°W direction on a falling slope of 23°. Find the clearance of the power line with the closest member of the frame. Show this clearance in all views.

9.3.16.   Scale: ¼ in. = 1 ft 0 in. Pipe clearance.

Y is 5 ft due south of X and 4 ft lower than X.

A level pipe runs due east from Y. From X a pipe runs S45°E on a downward slope of 40°.

1. What is the clearance between the centerlines of the two pipes?
2. Holding point X and the bearing of the pipe from X fixed, what will the new slope of this pipe have to be to have a centerline clearance of 24 in.?

Group 4. A line piercing a plane

9.4.1. Scale: 1 in. = 40 ft.

A, B, and *C* are three points of outcrop on a vein of ore, and *XZ* is a tunnel.

B is 50 ft east and 100 ft north of *A,* and 70 ft below *A.*

C is 130 ft east and 50 ft north of *A,* and 40 ft above *A.*

X is 80 ft due north of *A,* and 20 ft above *A.*

Z is 60 ft east and 10 ft north of *A,* and 40 ft below *A.*

Find where the tunnel pierces the vein, by three independent methods:

1. By using two views only and a vertical projecting plane.
2. By using two views only and a projecting plane showing as an edge in the front view.
3. By using the edge-view method.

9.4.2.[T] Scale: ¼ in. = 1 ft 0 in.

A rectangular-shaped bin with a sloping bottom, *ABCD,* which is pierced by a pipe.

Find where the centerline of the pipe pierces the bottom by three independent methods.

1. By using two views only and a vertical projecting plane.
2. By using two views only and a projecting plane showing as an edge in the front view.

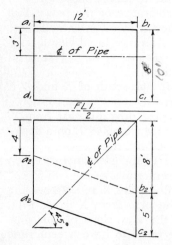

PROB. 9.4.2

3. By using the edge-view method.

9.4.3. Scale: 1 in. = 20 ft.

A is a point of outcrop on a vein of ore which dips down 45° toward the northeast. The upper and lower surfaces of the vein are parallel planes and the point *A* is on the upper plane. The strike of the vein is known to be N45°W. *B* is 40 ft east and 50 ft north of *A,* and 5 ft below *A.*

From *B* a shaft is driven bearing due west, and on a falling grade of 26.8 per cent. Measured along the centerline of this shaft, the distance through the vein is 20 ft. Find the real thickness of the vein.

9.4.4.[L] Scale: 1 in. = 100 ft.

A piece of sloping ground is represented by seven parallel straight-line contour lines bearing S15°E and spaced 50 ft apart on the map. The farthest one to the west is the 1,825-ft contour, the next one toward the east is the 1,850-ft line, the next one is the 1,875-ft line, etc., the ground being a uniformly sloping plane all the way.

A tunnel, 250 ft long, is driven from the point *A,* on the 1,825-ft contour, bearing N60°E on a falling slope of 45°. At what point, *C,* on the map would you start a second tunnel to bear N45°W from *C,* to fall 60°, and to meet the low end of the first tunnel? Locate *C* on the map by coordinates from *A.*

Find:

1. Number of feet *C* is south of *A.*
2. Number of feet *C* is east of *A.*
3. True length of the tunnel from *C.*

9.4.5.[L] Scale: 1 in. = 50 ft.

Problem 9.4.5 shows a few contour lines on the surface of the ground. They are shown as parallel straight lines, just to simplify the drawing layout. Assume the ground to be a plane between any two contour lines. From the point *A,* on the 1,900-ft contour line, the centerline of a road bears S30°W, and it is to have a constant falling grade of 20 per cent. It will cross the valley on a fill, and it is to tunnel through the hill.

1. To the scale given, draw the plan view and an elevation view looking due north (front elevation). Place these views in about the center of the sheet.

2. Locate the point *A* near the upper border and show the centerline of the road in the two given views.

3. Using only the plan and front views, find the points where the centerline of the road would enter the hill and come out again. Locate these points first by using a vertical cutting plane, and then independently by using a cutting plane appearing as an edge in the front view. See that both methods check.

4. Draw an edge view of the ground surface, locate the tunnel in this view, and check the piercing points again.

5. Scale the true length of the tunnel.

6. Scale the distance from *A* to the tunnel.

7. Scale the greatest depth of the fill over the 1,875-ft contour line, measured vertically.

9.4.6.   Scale: ⅛ in. = 1 ft 0 in.

Problem 9.4.6 shows two views of a portion of a steel tower on a flat roof. Using the *A* rays for light direction, find the shadow cast onto the flat roof by the two panels of the tower marked *A* and *B*. Show the shadow by dash lines.

9.4.7.   Scale: ⅛ in. = 1 ft 0 in.

Use the same tower and flat roof and the portion of the slanting roof as shown in Prob. 9.4.6. Find the shadow cast onto the flat and the slanting roofs by the panel *C* of the tower. Use the *B* rays for the direction of light and show the shadows by dash lines in both views.

9.4.8.   Scale: ⅛ in. = 1 ft 0 in.

A and *C* are the starting points of two pipes *AB* and *CD*, both bearing N45°E.

C is 28 ft due east of *A*.

B and *D* are 16 ft north of *A*.

B and *C* are at the same elevation, and *A* and *D* are at the same elevation 16 ft higher than *B* and *C*.

P is 22 ft east of *A* and 6 ft north of *A*, and 12 ft lower than *A*.

Using only the two given views, find the plan and elevation views of a single straight pipe connecting pipes *AB* and *CD* and point *P*. What is the bearing of the connecting pipe?

9.4.9.   Scale: ⅛ in. = 1 ft 0 in.

A store building with an inclined display window is shown in Prob. 9.4.9. A rifle bullet fired from *F*

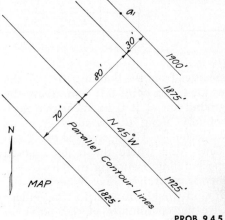

PROB. 9.4.5

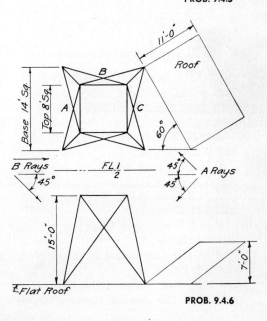

PROB. 9.4.6

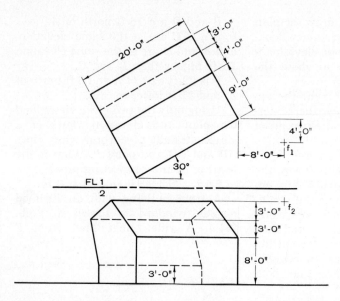

PROB. 9.4.9

emerges from the building through the window at a point 5 ft 0 in. above the ground and 3 ft 6 in. measured horizontally along the window from its western edge. Using only the two given views, find where the bullet entered the building assuming that its trajectory is a straight line. Where on the ground would you look for the bullet? Give the coordinates of this point on the ground with reference to the northwest corner of the building.

Group 5. Intersection of planes

9.5.1.[T]   Scale: ¼ in. = 1 ft 0 in. Two plane surfaces, ABC and XYZ, both to be considered indefinite in extent.

B is 10 ft west and 10 ft north of A, and 8 ft below A.

C is 15 ft west and 3 ft south of A, and 5 ft above A.

X is 8 ft west and 3 ft north of A, and 5 ft above A.

Y is 8 ft east and 10 ft north of A, and 3 ft below A.

Z is 4 ft east and 4 ft south of A, and 8 ft below A.

Using two views only, find four points on the line of intersection of the two planes. Find one point by each of the following methods.

1. By finding where the line XZ pierces the plane ABC.
2. By finding where the line AC pierces the plane XYZ.
3. By using any random vertical plane cutting both planes.
4. By using any cutting plane appearing as an edge in the front view. See that these four points check in a straight line in both views.

9.5.2.   Scale: ¼ in. = 1 ft 0 in. Two plane surfaces, ABC and DEF.

B is 5 ft north and 6 ft east of A, and 6 ft below A.

C is 2 ft south and 10 ft east of A, and 3 ft below A.

D is 2 ft south and 1 ft east of A, and 5 ft below A.

E is 7 ft north and 4 ft east of A, and 1 ft above A.

F is 4 ft north and 12 ft east of A, and 6 ft below A.

Show the line of intersection of the two planes in both views by using the edge view of one of the planes.

9.5.3.[T]   Scale: ⅜ in. = 1 ft 0 in. A, B, and C are three located points on a sloping ground.

B is 13 ft east and 2 ft north of *A*, and 2 ft above *A*.

C is 8 ft east and 9 ft south of *A*, and 4 ft below *A*.

The vertical centerline, *XY*, of a pier is 7 ft east and 2 ft 6 in. south of *A*. The upper pier surface is level, and is 4 ft above *A*. It is rectangular shaped and is 4 ft wide from east to west and 3 ft wide from north to south.

The east and west sides of the pier have a batter of 1 to 4. The north and south sides of the pier have a batter of 1 to 3. Using only the plan and front elevation, show only that part of the pier that is above the ground. Solve each view independently and check by projection.

9.5.4.   Scale: Actual size.

Draw the irregular-shaped pyramid shown in Prob. 9.5.4. Cut off the vertex of the pyramid by a plane in such a manner that the top of the remaining frustum will be a perfect parallelogram having two opposite sides each 1 in. long.

There are two possible solutions. (See Theorem 9, page 37.)

9.5.5.   Assume two planes to be located about as shown in Prob. 9.5.5. Their slopes are fixed but their relative elevations are immaterial for this problem. Plane No. 1 appears as an edge in elevation view 2 and plane No. 2 appears as an edge in elevation view 3. Both planes may be considered to be indefinite in extent.

1. Show the line of intersection of the two planes in all views.
2. Draw a new view which will show both planes as edges.

9.5.6.   Scale: Actual size.

Given two planes *ABC* and *ADE*, located as shown in Prob. 9.5.6, start on the right side of the sheet.

1. Find the line of intersection of the two planes.
2. Draw a new view of both planes showing the line of intersection as a point.

9.5.7.   Scale: Actual size.

The triangular pyramid shown in

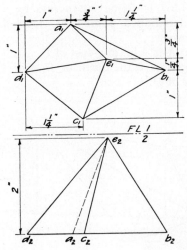

PROB. 9.5.4

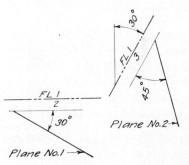

PROB. 9.5.5

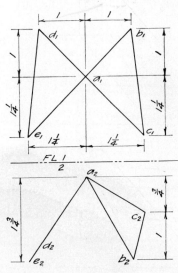

PROB. 9.5.6

Prob. 9.5.7 is to be cut across by a triangular-shaped plane ABC located as follows:

Point A is 4 in. due east of D.

Point B is 2 in. north of D and ½ in. east of D.

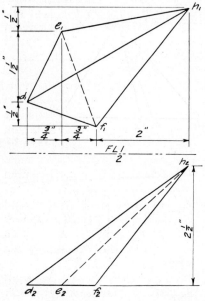

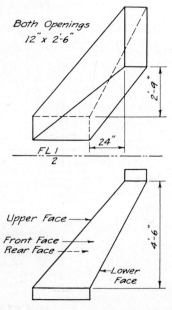

PROB. 9.5.7

Both Openings
12" x 2'-6"

PROB. 9.6.2

Point C is 2½ in. due west of F.

Point B is at the same elevation as the base.

Point C is 1 in. above the base, and point A is 1¾₆ in. above the base.

1. Using the two given views only, find the section cut out of the pyramid by the plane. Show it in both views and dash the hidden lines.

2. Check by using the edge view of the plane.

9.5.8.[T]   Scale: Actual size.

A rectangular block of metal 3 in. wide, 2 in. high, and 1¼ in. deep is cut by two planes ABC and ABD.

AB is a level line bearing N30°E passing over the center of the top of the block and ½ in. above it.

D is at the lower left rear corner of the block.

C is on the extended lower rear edge of the block 1 in. to the right of the lower right rear corner.

Draw the plan, front elevation, and right side elevation views of the block and planes and find the intersections of the planes and the block using only these views. Show the block in each view with the small cut-off portions discarded.

9.5.9.[T]   Scale: Actual size.

Use the same data as in Prob. 9.5.8, except draw the plan and elevation views only and solve the problem by drawing a new view that shows ABC and ABD as an edge.

Group 6. Dihedral angle

9.6.1.   Use the same wing wall and the same scale as for Prob. 9.1.1. Find the true size of the dihedral angle at the corner AB.

9.6.2.   Scale: ½ in. = 1 ft 0 in.

Offset piece made of steel plate.

Find the true size of the dihedral angle between the upper face and the front face.

9.6.3.   Use the same offset and the same scale as for Prob. 9.6.2.

Find the true size of the dihedral

angle between the rear face and the lower face.

9.6.4.[L]   Hopper made of steel plate. Layout scale: ¼ in. = 1 ft 0 in. Detail scale: 3 in. = 1 ft 0 in.

The dimensions shown in Prob. 9.6.4 are inside dimensions.

Adjacent side plates of the hopper are to be riveted to a special corner angle which is a bent plate and which is placed inside the hopper.

The problem is to detail the corner angle for the corner *AB*. Use a ⁵⁄₁₆- by 6-in. plate of length to suit, and bend it lengthwise in the middle to fit flush inside the hopper. The top and bottom ends are to be flush with the edges of the hopper. Use ½-in. rivets on about 4-in. centers on a 1¾-in. gage, and countersunk on the inside. Make a completely dimensioned steel-shop drawing of this angle ready for use, *i.e.*, bent, punched, and trimmed.

Method of procedure:

1. Layout work. On the left side of the sheet show the plan and the front elevation of the hopper, and solve for the dihedral angle at *AB* and the end cuts. Do this

in such a manner that these angles may be easily transferred to the shop drawing.

2. Mathematical work. Calculate the true length of the work line to the nearest ¹⁄₁₆ in. and express the bevel angles by their tangent with 12 as one leg (see Fig. 2.8).

3. Detailing work. On the right side of the sheet start the detail drawing, keeping it in the same position on the paper as the layout solution occupies, as in Prob. 9.6.4. Complete the detail drawing and show all necessary dimensions. Problem 9.6.4 shows a method of dimensioning which would be satisfactory to a steel-fabricating shop. The dimensions marked with a question mark must be determined. The dimensions marked *M*, *N*, *P*, and *R* must be such that the center of the end rivet is 1½ in. from the nearest sheared edge of the plate.

4. Tracing work. Trace the detail drawing only, squaring it up with the border as shown.

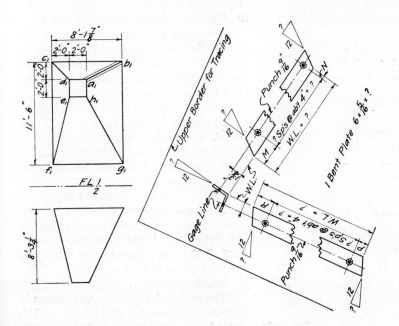

9.6.5.[L]   Use the same hopper, the same scales, and the same specifications as for Prob. 9.6.4. Make the solution for the corner CD.

9.6.6.[L]   Use the same hopper, the same scales, and the same specifications as for Prob. 9.6.4. Make the solution for the corner EF.

9.6.7.[L]   Use the same hopper, the same scales, and the same specifications as for Prob. 9.6.4. Make the solution for the corner GH.

9.6.8.[L]   Steel-plate hopper. The dimensions shown are outside dimensions.

Layout scale: ½ in. = 1 ft 0 in. Detail scale: 3 in. = 1 ft 0 in.

Make a completely dimensioned detail shop drawing of the riveted bent-plate connection between the plates for plane A and plane B.

The bent plate is to go on the outside of the hopper and is to fit flush at the top and bottom edges of the hopper.

Read Prob. 9.6.4 and follow the method of procedure suggested there.

9.6.9.[L]   Use the same data and the same scales as for Prob. 9.6.8. Make

the same solution for the plates between plane B and plane C.

9.6.10.[L]   Scale: ½ in. = 1 ft 0 in.

Use the hopper shown in Prob. 9.6.8.

Plane B is to be made of steel plate ⅟₁₆ in. thick and is to be flanged (or bent) at both side edges to fit against planes A and C. Flanges are to be 3 in. wide and must be trimmed so they will be flush with the top and bottom of the hopper.

Make a flat drawing of this flanged plate B and dimension it completely for the shop. Show the outside of the metal up.

9.6.11.[L]   Scale: ¼ in. = 1 ft 0 in.

Use the hopper shown in Prob. 9.6.4. The plane ABGH is to be made of steel plate ⅟₁₆ in. thick and is to be flanged (or bent) at both side edges to fit against the adjacent plates. Flanges are to be 3 in. wide and must be trimmed so they will be flush with the top and bottom edges of the hopper when assembled.

Make a flat drawing of this flanged plate and dimension it completely for the shop. Show the outside of the metal up.

9.6.12.[L]   Scale: ¼ in. = 1 ft 0 in.

Use the hopper shown in Prob. 9.6.4. Follow the same instructions as for Prob. 9.6.11 but make the solution and completely dimensioned flat drawing for the plane CDEF.

9.6.13.[L]   Scale: ¼ in. = 1 ft 0 in.

Use the hopper shown in Prob. 9.6.4. Follow the same instructions as for Prob. 9.6.11 but make the solution and completely dimensioned flat drawing for plane FGHE.

9.6.14.[L]   Scale: ¼ in. = 1 ft 0 in.

Use the hopper shown in Prob. 9.6.4. Follow the same instructions as for Prob. 9.6.11 but make the solution and completely dimensioned flat drawing for plane ABCD.

9.6.15.   Scale: 1½ in. = 1 ft 0 in.

The plan view of two steel beams is shown in Prob. 9.6.15. An 8- by 2¼- by 11.5-lb channel goes over the top

PROB. 9.6.8

of an 8- by 4- by 18.4-lb I beam. The webs of both beams are vertical planes, and the vertical clearance between the two beams must be ½ in. A bent-plate connection angle is to be fastened to the front face of the channel web and to the top flange of the I beam.

Find the angle (less than 90°) to bend the connection angle. Show an end view of each beam, assuming all metal to be ¼ in. thick.

9.6.16. In a ship, a vertical bulkhead (compartment wall) runs due north and south and is intersected by a steel plate. The top of the plate is cut off level and bears S60°W. The plate has a slope of 45° downward in a NW direction. What is the angle between the plate and bulkhead?

Group 7. Line perpendicular to a plane

9.7.1.[L] Scale: 1 in. = 100 ft.

A, B, and C are points of outcrop on a vein of ore.

A is 450 ft north and 280 ft west of B, and its elevation is 2,490 ft.

C is 110 ft north and 310 ft east of B, and its elevation is 2,280 ft.

D is 420 ft north and 150 ft east of B, and its elevation is 2,560 ft.

The elevation of B is 2,680 ft.

From the point D, which is not on the vein, it is desired to drive the shortest possible tunnel to the vein.

1. Draw the plan and front elevation views of the vein (neglecting any thickness), as located by the three points of outcrop.
2. In these same two views show the required tunnel and find where it hits the vein, using no other views.
3. Also, using no other views, find the shortest possible level tunnel and find where it hits the vein.
4. Check the piercing point of the shortest tunnel by locating it in an elevation view showing the vein as an edge.

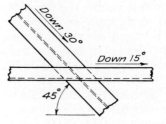

PROB. 9.6.15

5. Check the piercing point of the level tunnel by locating this tunnel in an inclined view showing the vein as an edge.
6. Scale the true length of each tunnel from D to the vein and the grade of the shortest tunnel.

9.7.2. Scale: ¼ in. = 1 ft 0 in. Plane surface, ABC.

A is 11 ft south and 9 ft east of B and 5 ft below B.

C is 8 ft south and 7 ft west of B and 10 ft below B.

A horizontal line, not on the plane, starts at C and runs due east for 8 ft. Project this line on to the plane ABC and show its projection on this plane in the two given views. Use only two views. Check by a right side elevation.

9.7.3. Use the same scale and the same data for the plane as in Prob. 9.7.2. A level line bears due south from B for a distance of 8 ft. Project this line on to the plane ABC and show its projection on this plane in the two given views. Use only two views. Check by a right side elevation.

9.7.4.[T] Use the same bin as in Prob. 9.4.2 and draw it to the same scale. Assume the bin to be closed at the top by a level cover. Show the centerline of a pipe which extends only between the bottom and the cover of the bin. This pipe is perpendicular to the bottom plane of the bin at its central point. Show this centerline in both views, using no other views. Mark plainly the upper end of the pipe in both views.

9.7.5. Use the same data and the same scale as for Prob. 9.3.7. Show where to connect the two given shafts with the shortest shaft. The location of the third shaft must be made by using only the two given views. A new view may be drawn to find the true length and true slope of the third shaft.

9.7.6. Use the same data and the same scale as for Prob. 9.3.5. Show where to connect the two given tunnels with the shortest possible shaft, using only the two given views. After this shaft has been located, a new view

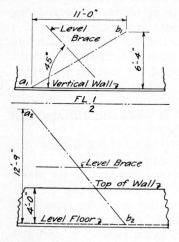

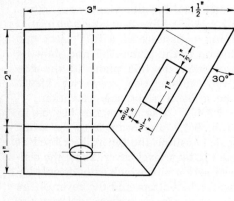

PROB. 9.7.8

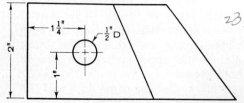

PROB. 9.7.10

may be drawn to find its true length and true slope.

9.7.7.[L] Scale: ¼ in. = 1 ft 0 in. Plane surface, *ABC*.

B is 4 ft south and 8 ft east of *A* and 9 ft above *A*.

C is 9 ft south and 4 ft west of *A* and 5 ft above *A*.

1. At a point on the plane equidistant from *A*, *B*, and *C* erect a perpendicular above the plane and make it 6 ft long. Show it in all views.
2. Locate this perpendicular by projection in an inclined view showing the plane as an edge. See that it checks at right angles to the plane.

9.7.8. Scale: ¼ in. = 1 ft 0 in.

Problem 9.7.8 shows a brace, *AB*, which is intersected at its center point by a level brace. From the point of intersection of these two braces it is desired to run a strut perpendicular to the plane of the two braces.

1. Using only the two given views, show the strut in these views.
2. Find how far below the top of the wall the strut would be fastened to the wall, using only two views.
3. Using a third view, find the true length of the strut.

9.7.9. Scale: 1 in. = 1 ft 0 in.

AB is a steel bar. Point *B* is 18 in. east and 24 in. north of *A* and 24 in. lower than *A*. Point *X* is 18 in. due north of *A* and 20 in. lower than *A*.

A rectangular plate is perpendicular to *AB*, it has one corner at *X*, and two of its edges are level. The bar passes through the center of the plate.

Using two views only, show this plate in both views. Check by using extra views. Show the plate in its true size and measure it.

9.7.10. Scale: Actual size.

A rectangular hole is to be broached through the steel block shown in Prob. 9.7.10 starting with the given opening in the slant face and in a direction perpendicular to the slant face. Show the complete hole in both views using only the two given views. What is the clearance or interference with the drilled hole?

Group 8.   Angle between a line and a plane

9.8.1.[L]   Solution scale: ½ in. = 1 ft 0 in. Detail scale: 3 in. = 1 ft 0 in.

Problem 9.8.1 shows two views of a plane to which a casting must be fastened for anchoring a strut from the point X. The centerline of the strut is as shown and it slopes down 20° (tangent 20° = 0.364) as it leaves X. It is not necessary to draw the front elevation.

1. Find the centerline length of the strut from X to where it meets the plane.
2. Find the angle N, in degrees, which the strut makes with the plane.
3. Find the angle M, in degrees, which the centerline appears to be away from the 6-in. side of the casting base, when looking perpendicular to the base. The 6-in. side is to be parallel to AB on the plane.
4. If it is desired to make the angle M = 0°, find the slope of the strut. Keep the point X fixed and the direction of the strut in the plan the same.
5. Make a detailed drawing of a casting similar to the one shown, for receiving the strut in question 3 above. The base of the casting is 4 by 6 in., and the metal is ¼ in. thick all over. The receiving socket should have a hole ½ in. in diameter with a clear depth of 3 in.

9.8.2.[T]   Scale: 1 in. = 1 ft 0 in.

Problem 9.8.2 shows an approximation of a portion of a wing on an airplane to which the strut AC is to be fastened at C. Design the necessary fitting for receiving the strut similar to the one shown in Prob. 9.8.1. The dashed lines in the plan indicate how

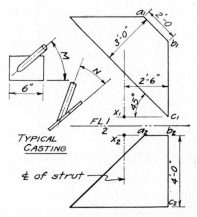

PROB. 9.8.1

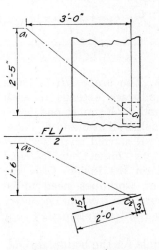

PROB. 9.8.2

the base of the casting is to be placed. Make the solution for the angles M and N.

9.8.3.[L]   Scale: ⅛ in. = 1 ft 0 in.

Steel stack running up the outside of a building and braced to a vertical wall. Problem 9.8.3 shows the plan view of an actual installation which was recently made at a Seattle plant.

The points A, B, C, E, F, and G are on the inside top edge of the vertical wall. The points M and N are on a steel ring 2 ft below the top of the stack. The top of the stack is 30 ft 2 in. above the top of the wall. The points D and H are the center points of the long braces AM and GN, respectively, whose centerlines intersect the center

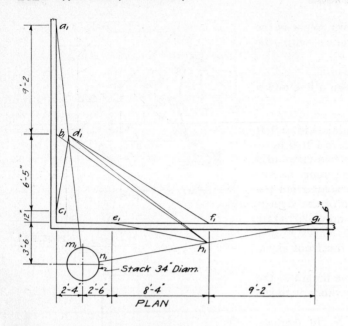

of the stack. The brace DH is level. From D one brace extends to each wall at C and F. From H one brace extends to each wall at B and E. These braces were all starred angle sections with bent-plate connections fastened to the wall on the inside. Only the centerlines or worklines need to be dealt with here and it is not necessary to draw any front elevation.

Find:

1. The angle for the bent-plate connection at A.

2. The angle for the bent-plate connection at B.

3. The angle for the bent-plate connection at C.

9.8.4.[L]   Use the same data and scale as for Prob. 9.8.3.

Find:

1. The angle for the bent-plate connection at E.

2. The angle for the bent-plate connection at F.

3. The angle for the bent-plate connection at G.

9.8.5.   Scale: ¼ in. = 1 ft 0 in.

An A frame, ABC, and two guy wires fastened to the top of point C.

Find the true size of the angle between the guy CE and the plane of the frame.

9.8.6.   Use the same data and scale as for Prob. 9.8.5.

Find the true size of the angle between the guy CD and the plane of the frame.

9.8.7.   Scale: ¼ in. = 1 ft 0 in.

From the point X a ray of light goes in the direction shown in both views in Prob. 9.8.7, and it is reflected from the metal roof.

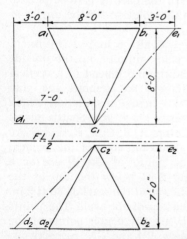

PROB. 9.8.5

1. Using only the two given views find the reflected ray in both views.
2. Check by extra views.
3. Find the angle the ray makes with the roof.

9.8.8.   Scale: ⅛ in. = 1 ft 0 in.

Point A is located on the surface of a vertical wall which bears N45°E. A horizontal steel beam 18 ft long projects due north from A terminating in point B. The beam is braced at B from two points in the wall, X and Y.

X is 14 ft east of B and 12 ft higher than B.

Y is 14 ft east of B and 6 ft lower than B.

Find the lengths of braces BX and BY and the angles they make with the wall.

Group 9. Revolution

9.9.1.   Scale: Actual size.

A level line AB is 3 in. long and bears S45°E from A.

Point X is 2 in. due south of A and 1 in. higher than A.

Revolve the point X about the line AB as an axis until it lies at an elevation ½ in. lower than AB. Call the new position of the point X' and show it in the plan and the front elevation. There are two solutions.

9.9.2.ᴸ   Scale: ½ in. = 1 ft 0 in.

AB is the centerline of a shaft which makes an acute angle with the vertical side wall and the level base of a machine. A crank 27 in. long is at right angles to the shaft AB at the midpoint of the shaft.

1. In the plan and the front elevation show the crank in the following positions with dash lines:
   a. As it hits the base.
   b. As it hits the side wall.
   c. When it occupies its highest position.
2. What is the maximum number of degrees through which the crank can rotate?
3. What is the maximum length the

crank could be and just clear the wall?
4. What is the maximum length the crank could be and just clear the floor?
5. How far will the side wall have to be moved back to allow the 27-in. crank to rotate through exactly 180°?

Tabulate the results for questions 2, 3, 4, and 5 and indicate clearly on the drawing where each result was measured.

9.9.3.ᴸ   Scale: ½ in. = 1 ft 0 in.

AB is the centerline of a shaft which operates a control valve on a

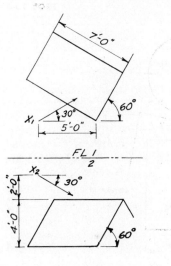

PROB. 9.8.7

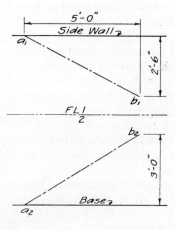

PROB. 9.9.2

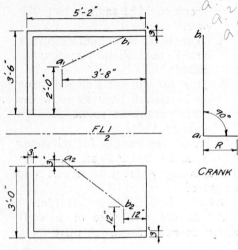

PROB. 9.9.3

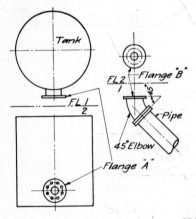

PROB. 9.9.4

9.9.4. A steel tank with a flanged elbow connection.

Flange B on the elbow is to have bolt holes drilled in it so that, when it is bolted to flange A on the tank, the straight pipe will have a true grade of 66.7 per cent downward to the right.

Flange A is 1 in. thick and 6 by 11 in. in diameter, and has eight ⅞-in. holes on a 9½-in. bolt circle.

It is standard practice for the bolt holes to straddle the centerline as shown on flange A. It will be necessary to locate on flange B the centerline which the holes will straddle so that the given condition will be satisfied. The elbow is a standard 45° elbow with an undrilled flange.

1. Find the angle between the elbow centerlines as shown and the centerlines to be used for drilling bolt holes.
2. Find the bearing of the pipe after it is installed, assuming that flange A faces south.
3. Complete the drawing of flange B and elbow, scale 1½ in. = 1 ft 0 in., showing the holes properly located.

Note: It is not necessary to draw the tank. Use centerlines only.

9.9.5. Scale: ¼ in. = 1 ft 0 in. plane surface ABC.

B is 4 ft due north of A and 4 ft below A.

C is 6 ft north and 8 ft east of A and 8 ft below A.

1. Revolve the plane ABC until it shows in its true size in the plan.
2. Revolve the plane ABC until it shows in its true size in the front elevation.

9.9.6.[T] Scale: 1½ in. = 1 ft 0 in.

AB is the centerline of a rod passing through the disk which is hinged on the left side as shown. The disk can move forward only through an angle of 120°. Show the centerline of the slot which must be cut in the disk to allow it to have this motion and show it only on the front face of the disk.

9.9.7. Scale: ¼ in. = 1 ft 0 in.

machine. A portion of the frame of the machine is shown. This shaft is turned by a 90° handle fastened to the shaft at A.

1. Find the longest possible handle length, R, which will permit the shaft to be turned through a complete revolution without interfering with the frame.
2. There are three places to be investigated. Show all three interference points in the two given views.
3. Using the maximum handle length, show the path of the outer end of the handle in the two given views.

A, B, and *C* are points of outcrop on a body of ore.

B is 10 ft north and 5 ft east of *A* and 7 ft below *A*.

C is 5 ft north and 13 ft east of *A* and 4 ft above *A*.

Find the true dip of the vein, using only the plan and front elevation views.

9.9.8.   Scale: 1 in. = 40 ft.

A, B, and *C* are three located points on a plot of ground.

B is 100 ft north and 50 ft east of *A*, and 70 ft below *A*.

C is 50 ft north and 130 ft east of *A*, and 40 ft above *A*.

Find the centerline of a road from *A* having a 20 per cent rising grade. Show it in the plan and front elevation, using only these two views.

9.9.9.   Scale: ¼ in. = 1 ft 0 in.

A level pipe bearing due south projects 1 ft out from a vertical wall running east and west, and is 8 ft above a level floor. A one-sixth bend (120° angle) is screwed on to the end of this pipe. A pipe screwed into this bend makes an angle of 40° with the floor and just touches the floor. Using only the plan and the front elevation, show the pipe in these two views.

Find:

1. The true length of the pipe.
2. The distance the lower end of the pipe is out from the wall.

9.9.10.   Use the same data and the same scale as for Prob. 9.8.5. By the method of revolution find the true size of the angle the guy *CE* makes with the plane of the frame.

9.9.11.   Use the data and the scale for the shaft *AB* in Prob. 9.9.2.

Divide a small sheet into two parts by a line parallel to the short edge of the paper. Draw two views of the shaft only in each half of the sheet.

In one half of the sheet revolve the shaft *AB* until it shows in its true length in the plan.

In the other half of the sheet revolve the shaft *AB* until it shows in its true length in the front elevation.

Show and label the axis of revolution used in all four views.

See that the true lengths obtained by both methods check.

Mark the true slope of the shaft where it is seen.

9.9.12.   Use the same data and the same scale as for Prob. 9.6.2. By the method of revolution find the true size of the dihedral angle between the upper face and the rear face.

9.9.13.   Scale: 1 in. = 1 ft 0 in.

AB is the centerline of a level pipe which is to be bent at *B* so as to connect with the point *C*. Point *C* is at an elevation 26 in. higher than *AB*. In pipe shops this is called a rolling offset.

1. Find in degrees the true size of the angle *M* to which the pipe must be bent.

2. Measure the true length of the portion of the pipe *BC*.

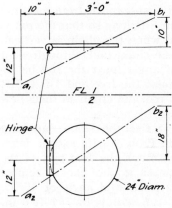

PROB. 9.9.6

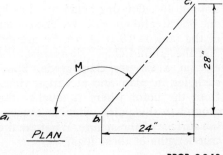

PROB. 9.9.13

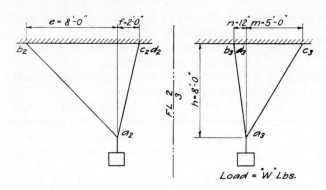

9.9.14. Use the same data and scale as for Prob. 9.5.6. Revolve the plane ADE until the dihedral angle between the two planes at A is exactly 45°. The line of intersection of the two planes must remain unchanged during the revolution.

9.9.15. Scale: ¼ in. = 1 ft 0 in.

A steel rod 10 ft long has one end attached to a vertical north-south wall with a ball-and-socket joint and the other end is suspended from a ceiling by a chain 8 ft long so that the rod is horizontal when the chain is vertical. Keeping the chain taut, through what horizontal angle must the rod be rotated to raise the free end 4 ft 0 in.? Show the rod and chain in their new positions.

9.9.16. Use the same data and scale as in Prob. 9.9.15, except determine how much the free end of the bar is raised if the bar is rotated through a horizontal angle of 45°.

9.9.17.[T] Scale: ⅛ in. = 1 ft 0 in.

A 20-ft-diameter spherical tank is supported by three radial legs, each sloping 45°. One leg bears due north, one S60°E, and one S60°W. The center of the tank is 18 ft above the ground and the ground slopes downward due east at a 20 per cent grade. Show each leg in plan and elevation views from the point where it touches the ground to the point where it touches the surface of the sphere. What are the true lengths of these legs?

Group 10. Noncoplanar structures and vectors

The problems in this group are all to be solved by the graphical method. Hard and well-sharpened pencils must be used in order to obtain accurate results.

9.10.1.[L] Space scale: ½ in. = 1 ft 0 in. Force scale: To be selected.

A hanging structural-steel frame of three members which support a given load. Find the load on each member.

9.10.2.[L] Use the same scales and the same frame as for Prob. 9.10.1 but substitute the following values in the space drawing.

$$h = 10 \text{ ft } 0 \text{ in.}$$
$$e = 6 \text{ ft } 0 \text{ in.}$$
$$f = 2 \text{ ft } 0 \text{ in.}$$
$$n = 2 \text{ ft } 0 \text{ in.}$$
$$m = 5 \text{ ft } 0 \text{ in.}$$

Solve for the load on each member.

9.10.3.[L] Use the same scales and the same frame as for Prob. 9.10.1 but substitute the following values in the space drawing.

$$h = 8 \text{ ft } 0 \text{ in.}$$
$$e = 6 \text{ ft } 0 \text{ in.}$$
$$f = 3 \text{ ft } 0 \text{ in.}$$
$$n = 4 \text{ ft } 0 \text{ in.}$$
$$m = 2 \text{ ft } 0 \text{ in.}$$

Solve for the load on each member.

9.10.4.[L]   Space scale: 1 in. = 1 ft 0 in. Force scale: To be selected.

A steel-frame tripod of three members and supporting a given load. Find the load on each member which is caused by the given load to be selected.

9.10.5.[L]   Space scale: 3/16 in. = 1 ft 0 in. Force scale: To be selected.

A derrick with a boom elevated at an angle of 30° and supporting a given load at its end. Load to be selected.

Find the load on:
1. The boom *AC*.
2. The tie *BC*.
3. The mast *AB*.
4. The two guys *BD* and *BE*.

9.10.6.[L]   Space scale: 1 in. = 10 in. Force scale: 1 in. = 400 lb.

The cabane system of support between the wing and the fuselage of an airplane. A known horizontal load and a known vertical load are acting at *A*.

Draw the plan and front elevation of the three members and the loads. Draw the vector diagram for the five noncoplanar forces acting at *A*.

Find the value of the force acting in each of the three members *AB*, *AC*, and *AD*. Indicate whether each load is a tension or a compression load.

9.10.7.   Vector scale: 1 in. = 10 ft per second.

Three vectors representing velocities have the directions in space as shown in Prob. 9.10.7. The values of these velocities are as follows:

$$DA = 24 \text{ ft per second}$$
$$DB = 18 \text{ ft per second}$$
$$DC = 12 \text{ ft per second}$$

Using only the two given views draw the vector diagram and find:
1. The value of the resultant velocity.
2. The true slope of the resultant velocity.
3. The direction of the resultant velocity in the plan. Give the angle between its plan view and d_1a_1.

9.10.8.   Vector scale: 1 in. = 30 ft per minute.

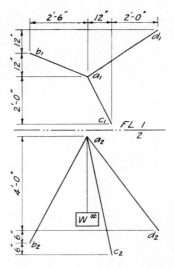

PROB. 9.10.4

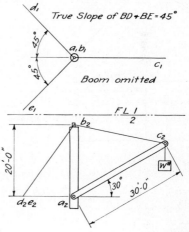

PROB. 9.10.5

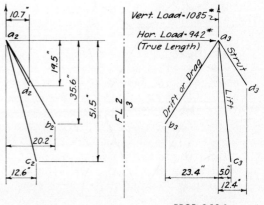

PROB. 9.10.6

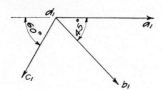

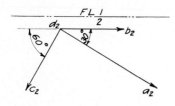

PROB. 9.10.7

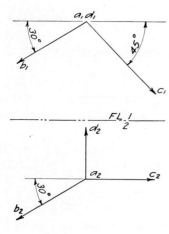

PROB. 9.10.8

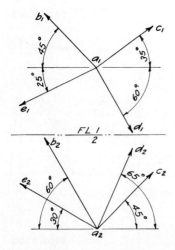

PROB. 9.10.9

Three velocity vectors are fixed in space as shown in Prob. 9.10.8. The values of the velocities are:

$$AD = 60 \text{ ft per minute}$$
$$AC = 113 \text{ ft per minute}$$
$$AB = 134 \text{ ft per minute}$$

Using only the two given views draw the vector diagram and find:

1. The value of the resultant velocity.
2. The slope of the resultant velocity.
3. The direction of the resultant velocity in the plan. Give the angle from a_1b_1 or a_1c_1.

9.10.9.[T]   Vector scale: 1 in. = 20 ft per second.

Four velocity vectors are fixed in space as shown in the sketch. The true values of the four velocities are:

$$AB = 50 \text{ ft per second}$$
$$AC = 38.5 \text{ ft per second}$$
$$AD = 47 \text{ ft per second}$$
$$AE = 43 \text{ ft per second}$$

Using only the given views draw the vector diagram and find:

1. The value of the resultant velocity.
2. The slope of the resultant velocity.
3. The direction of the resultant velocity in the plan referred to a_1c_1.

9.10.10.[L]   Use the same scale and the same data as for Prob. 9.10.4. On a large sheet make the solution by the method of Section 5.15.

9.10.11.[L]   Use the same scale and the same data as for Prob. 9.10.1. On a large sheet make the solution by the method of Section 5.15.

9.10.12.[L]   Space scale: ¾ in. = 1 ft 0 in. Force scale: 1 in. = 200 lb.

Given a tripod frame with one load acting at the pin F as shown. Find graphically the load on each of the three members, using the line-as-a-point method.

9.10.13.[L]   Use the same tripod frame and the same scales as for Prob. 9.10.12, but omit the load.

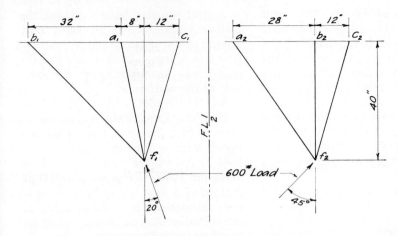

PROB. 9.10.12

Find the value of a load applied at F and the direction of its line of action to produce a stress of 600 lb compression in each of the three members.

9.10.14. Space scale: 1 in. = 20 ft. Force scale: 1 in. = 400 lb.

A vertical mast 40 ft tall is supported by two cables attached to the top of the mast and anchored in the ground at the same level as the bottom of the mast at points A and B.

A is 25 ft north and 30 ft west of the mast.

B is 20 ft south and 30 ft west of the mast.

A force of 1,000 lb acts horizontally at a bearing of S75°E at the top of the mast pulling toward the east. What are the loads on the cables and mast?

9.10.15. Vector scale: 1 in. = 10 amp.

Vector geometry is extremely useful in analyzing alternating-current electric circuits. In a circuit with three parallel branches the total current is the vector sum of the coplanar vectors shown in Prob. 9.10.15. The values of I_1, I_2, and I_3 are 19, 15, and 24 amp, respectively. Find the value of the total current and the angle (phase difference) by which it leads or lags the voltage E. Leading angles are measured counterclockwise.

9.10.16. Space scale: 1 in. = 2 nautical miles. Relative motion.

Ship B is 11 nautical miles east of A and 4 nautical miles north of A. Ship A is steering a N45°E course at a speed of 20 knots, and ship B is on a N60°W course at 16 knots. At their closest position, how far apart are the two ships? How far did each ship travel to reach this position? How many minutes did this take? (*Hint:* Hold one ship fixed and determine the relative motion of the other ship to the fixed ship.)

Group 11. Cylinders

9.11.1. Scale: Actual size. An oblique cylinder and a line AB.

Draw an oblique cylinder having an axis 3 in. long (true length), bearing

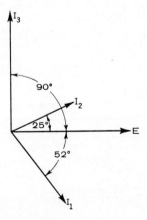

PROB. 9.10.15

N45°W and sloping up 45° (true slope). Both upper and lower bases are level circles 1½ in. in diameter.

Line *AB*. *A* is 1½ in. due west and ¾ in. above the center of the lower base. *B* is 1¾ in. east and 1½ in. north of *A*, and 1⅜ in. above *A*.

Using two views only, find the two points where the line *AB* pierces the cylinder. Check these points by drawing a view showing the cylinder as an edge.

9.11.2. Scale: ¼ in. = 1 ft 0 in.

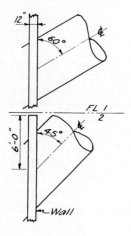

PROB. 9.11.2

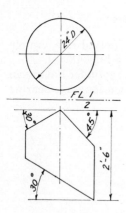

PROB. 9.11.5

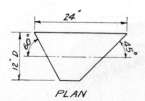

PLAN

PROB. 9.11.6

A steel penstock passes through the wall of a powerhouse. The outside diameter of the penstock is 6 ft. Find the true size of the opening to allow in the wall. Show the opening on the near side of the wall only.

9.11.3. Scale: ¼ in. = 1 ft 0 in.

Use the same data as shown in Prob. 9.11.2 except for the pipe size which will be different. Suppose the hole in the wall is round and 6 ft in diameter. Find the true shape the right section of a pipe would have to be to pass through the hole. Show a line on the developed pipe where a flange could be welded to the pipe so that the flange would bear against the wall all the way around.

9.11.4.[T] Scale: 1½ in. = 1 ft 0 in.

From a given point, *A*, the center-line of a flanged pipe bears N45°E and rises with a true slope of 30°. The pipe has an inside diameter of 12 in., an outside diameter of 13 in., and an over-all length of 16 in. On the far end is a flange ¾ in. thick and 17 in. in diameter.

Show only the visible portion of the pipe in the plan and front elevation views. Show only four 1³⁄₁₆-in. bolt holes through the flange and spaced 90° apart around the flange on a bolt circle 15 in. in diameter.

9.11.5. Scale: 1 in. = 1 ft 0 in.

A steel cylindrical shell has three cuts across its surface as shown in Prob. 9.11.5.

Make the complete development of the portion of the shell shown.

9.11.6. Scale: 1½ in. = 1 ft 0 in.

A cylindrical pipe is shown having two angular cuts at the ends. Develop the portion of the pipe which is shown.

9.11.7.[L] Scale: 1 in. = 1 ft 0 in.

The line *AB* is a portion of the centerline of a pipe 18 in. in diameter. The point *A* lies in the level floor and the point *B* lies in the vertical wall.

1. Show the plan and front elevation of the portion of the pipe which lies between the floor and the wall.

2. Show the development of a piece of pipe covering to cover the portion of the pipe lying between the floor and the wall.

3. Assuming the pipe is to be cut from a stock length, find the angle of twist between the two end cuts.

9.11.8.[L]   Use the same scale and the same layout for the line AB as for Prob. 9.11.7. Let the pipe be only 15 in. in diameter and let the floor be inclined with a rising 20 per cent grade in going from A to the wall. Point A lies in this inclined floor and point B is still the same height above A.

Solve questions 1, 2, and 3 of Prob. 9.11.7 for this changed condition.

9.11.9.[T]   Scale: 3 in. = 1 ft 0 in.

The line MN is the axis of a cylinder which is shown in the front elevation only in the sketch.

Using only these two given views, find the two points where the line AB pierces the cylinder.

Check these two points by the edge-view method.

9.11.10.   Scale: 3 in. = 1 ft 0 in.

Draw the plan and the front elevation of the cylinder and the point A as they are located in Prob. 9.11.9.

1. In the two given views show two planes, both of which contain the point A and are tangent to the cylinder.

2. Find the true slope of both of these tangent planes.

Group 12. Cones

9.12.1.   Scale: Actual size.

Draw a right cone of revolution with a base 3 in. in diameter and a 3½-in. altitude. The axis is vertical. The vertex of the cone is V.

Line AB. A is 2 in. due west of V and 2¼ in. below V. B is 1 in. east and ½ in. north of V and 1 in. below V.

Using only the plan and front elevation views, find the two points where the line AB pierces the cone and show them in both views.

9.12.2.[L]   Scale: Actual size. Refer to Fig. 6.10.

Assume the cone in Fig. 6.10 to have a vertical axis 3½ in. long and a level base 3 in. in diameter. Cut this

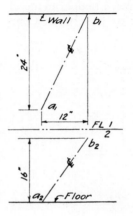

PROB. 9.11.7

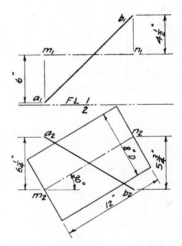

PROB. 9.11.9

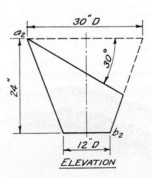

ELEVATION

PROB. 9.12.3

cone by the four planes B, C, D, and E as shown, assuming the plane B to be 1½ in. below the vertex of the cone.

Show each cut across the cone in the plan view and also in its true size.

9.12.3.   Scale: 1 in. = 1 ft 0 in.

The sketch shows a hopper made of steel plate. The axis is vertical. Com-

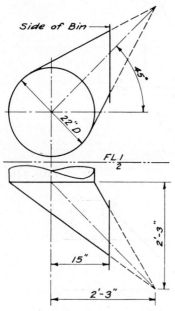

PROB. 9.12.4

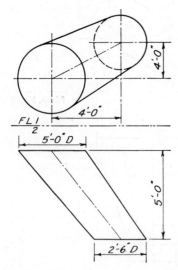

PROB. 9.12.5

plete the plan view and show the development of the plate. Also show in the elevation view the shortest path on the surface from a_2 to b_2.

9.12.4.[L]   Scale: 1½ in. = 1 ft 0 in.

Problem 9.12.4 shows a hopper made of steel plate and opening into the vertical side wall of a bin. Develop the plate for the hopper. Show the true shape of the hole to be cut in the side wall of the bin.

9.12.5.[L]   Scale: ½ in. = 1 ft 0 in.

A conical-shaped hopper connecting two round pipes. Develop the steel plate for making this hopper.

9.12.6.   Scale: 1 in. = 1 ft 0 in.

The sketch shows a hopper made of steel plate and opening into the vertical side wall of a bin. Develop the plate for the hopper. Show the true shape of the hole to be cut in the side wall of the bin.

9.12.7.   Scale: ¾ in. = 1 ft 0 in.

A given cone of revolution with a level axis and a point X not on the cone.

1. Show the two possible planes which may be drawn tangent to the cone and containing the point X.

2. Find the true slope of each of these tangent planes.

9.12.8.   Scale: ¼ in. = 1 ft 0 in.

Draw a right cone of revolution having a vertical axis 12 ft high and a base 12 ft in diameter.

Locate a point X so that it is 12 ft east, 1 ft south, and 6 ft below the vertex of the cone.

Using only the plan and front elevation views, show a plane which is tangent to the cone and which contains the point X. There are two possible solutions.

9.12.9.[T]   Scale: ⅛ in. = 1 ft 0 in.

A line AB is the centerline of a pipe line. A point C is not on the line AB.

B is 18 ft east and 10 ft south of A and 19 ft below A.

C is 6 ft east and 19 ft south of A and 19 ft below A.

1. Show how to connect *C* with the pipe *AB*, using a pipe having the same grade as the pipe *AB*.
2. Find the true length of this connecting pipe.

Group 13. Spheres

9.13.1.   Scale: Actual size.

Draw any three orthographic views of a sphere 3 in. in diameter. Point *C* is the center of the sphere and points *A* and *B* are two points on the surface of the sphere. In the plan view, *A* is ¾ in. to the left and ½ in. in front of *C*. Also the point *B* is 1¼ in. in front and ½ in. to the right of *C*.

Show both points *A* and *B* on the sphere in all three views.

Show the true length of the shortest distance from *A* to *B* measured on the surface of the sphere, using only the two given views.

9.13.2.   Scale: Actual size.

Draw the plan and front elevation views of a sphere 2 in. in diameter. The center of the sphere is *C*. Locate the line *AB*.

A is 1¼ in. south and ¼ in. west of *C*, and 1½ in. above *C*.

B is 1 in. east and 1 in. north of *C*, and ¾ in. below *C*.

Find the two points where the line *AB* pierces the sphere.

1. By the great-circle method.
2. By the small-circle method.

9.13.3.   Scale: Actual size.

Three spheres, whose diameters are 1½ in., 1¾ in., and 2 in., respectively, have their centers in the same level plane. They are also tangent to each other. A fourth sphere, whose diameter is 2½ in., is tangent to all three of the given spheres.

Show all four spheres in the plan and in a side elevation view.

Dash the invisible part of each sphere.

9.13.4.   Scale: Actual size.

Three spheres, whose diameters are 1½ in., 1¾ in., and 2 in., respectively,

are tangent to each other and resting on a level plane. A fourth sphere 2½ in. in diameter rests on top of the other three spheres and is tangent to them all. Place the smallest sphere in front.

Show all four spheres in the plan and in the right side elevation view.

Dash the invisible part of each sphere.

9.13.5.   Scale: ¼ in. = 1 ft 0 in.

A steel water tank, supported on a tower, has a hemispherical bottom 15 ft in diameter. Show the approximate development for one of the steel plates for the bottom, using the meridian

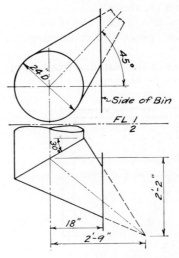

PROB. 9.12.6

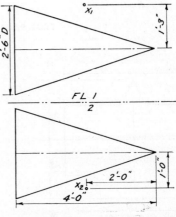

PROB. 9.12.7

method. Divide the bottom into 12 sections.

9.13.6.   Scale: ⅛ in. = 1 ft 0 in.

The metal dome of a building is to be made hemispherical in shape and 30 ft in diameter. Show the approximate development of the steel plates, using the zone method. Divide the dome into six sections and assume that the plates are not over 24 ft long.

9.13.7.   Scale: 3 in. = 1 ft 0 in.

A line *AB* is given and also a sphere 10 in. in diameter whose center is at *C*.

A is 5 in. to the right and 6 in. back of *C*, and 4 in. above *C*.

B is 9 in. to the right and 0 in. back of *C*, and 8 in. above *C*.

It is possible to have two planes tangent to the sphere and containing the line *AB*.

1. Show in the plan and front eleva-

tion these two tangent planes and their points of tangency with the sphere.
2. Find the size of the dihedral angle in degrees between these two tangent planes.
3. Find the true slope of each plane.

Group 14. Intersections

9.14.1.   Scale: ½ in. = 1 ft 0 in.
Conical reducer.

The cylinder and the cone are both surfaces of revolution. Their axes are both level but they do not intersect. The axis of the cone is 3 in. below the axis of the cylinder.

1. Show the complete curve of intersection in the plan.
2. On a separate sheet develop the conical and the cylindrical surfaces.

9.14.2.   Scale: ½ in. = 1 ft 0 in.
In Prob. 9.14.1 let the two axes be at the same level so they intersect.

1. Show the complete curve of intersection in the plan using the sphere method.
2. On a separate sheet develop the conical and the cylindrical surfaces.

9.14.3.   Scale: Actual size.
A cone of revolution and a cylinder of revolution intersecting. The axes are both in their true length in the view shown but they do not intersect. The axis of the cone is ⅝ in. behind the axis of the cylinder.

1. Show the complete curve of intersection in the given view and in the plan.
2. On a separate sheet develop the conical and the cylindrical surfaces.

9.14.4.   Use the same scale and the same data as for Prob. 9.14.3, except that the axes are to be taken as intersecting.

1. Show the complete curve of intersection in the given view and also in the plan, using the sphere method.

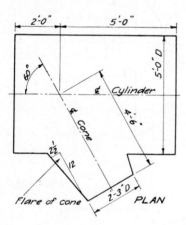

PROB. 9.14.1

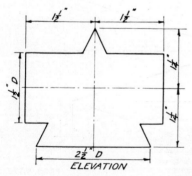

PROB. 9.14.3

2. On a separate sheet develop the conical and the cylindrical surfaces.

9.14.5. Scale: ¾ in. = 1 ft 0 in.

A concrete wing wall as shown must have a hole left in it large enough to admit a level pipe having a 24-in. outside diameter.

The line X to Y is a straight line. The front face through which the hole passes is the only slanting face on the wing wall. The front elevation is complete as shown.

Draw complete plan, front elevation, and right side elevation views.

9.14.6. Use the same scale and the same data as for Prob. 9.14.5, except that the line from X to Y is a part of a circle having its center directly in front of X and on the same level as X.

Draw complete plan, front elevation, and right side elevation views.

9.14.7.[L] Scale: 1 in. = 1 ft 0 in.

Two intersecting ventilating pipes as shown in Prob. 9.14.7.

1. Complete the curve of intersection of the two pipes in the front elevation.
2. In both the given views, show the outer end of the 20-in. pipe as it would appear if cut off square with its axis and at a distance of 2 ft 7 in. from the point A.
3. Develop the surfaces of both pipes.

9.14.8.[L] Scale: 1½ in. = 1 ft 0 in.

Double conical offset.

The three vertical ventilating pipes shown in part are to be connected by a double conical offset. Draw the given views, locating the center of the 20-in. pipe 4 in. from the left border and ¼ in. from the lower border and 1½ in. from the upper border.

1. Show the complete line of intersection of the two cones. Show all hidden lines dashed and mark all tangent points accurately.
2. Develop the small cone. Show the outside of the metal up and make the seam as short as possible.

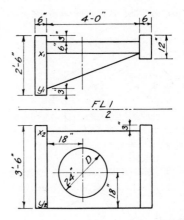

PROB. 9.14.5

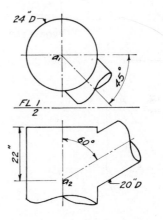

PROB. 9.14.7

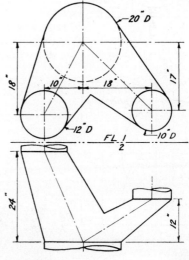

PROB. 9.14.8

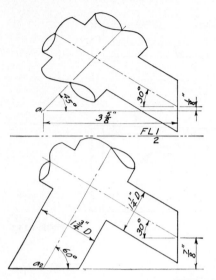

PROB. 9.14.9

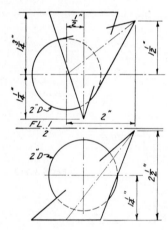

PROB. 9.14.11

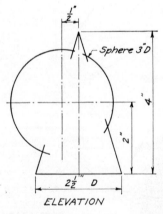

ELEVATION

PROB. 9.14.12

9.14.9.[L]   Scale: Actual size.

Two intersecting pipes have their bases in different planes. Both pipes are cylinders of revolution.

Show the entire curve of intersection in the two given views. Do not try to develop the surfaces from these views.

9.14.10.[L]   Use the same data and scale as in Prob. 9.14.9.

Start on the left. Obtain a view which shows both pipes in their true lengths. In this view show only the intersection of the two pipes. It is not necessary to show the ends of either pipe.

9.14.11.   Scale: Actual size.

Two intersecting cones having their bases in different planes. One of the cones is a cone of revolution with a level axis.

1. Show the entire curve of intersection in both the given views.
2. Develop both the cones.

9.14.12.   Scale: Actual size.

A sphere intersected by a cone of revolution. The axis of the cone is vertical; it is ½ in. to the right and ½ in. behind the center of the sphere.

1. Show the entire curve of intersection in both the plan and front elevation. Use both circular and straight-line elements on the cone in determining points. (Two different methods.)
2. Develop the cone.

9.14.13.   Scale: ¾ in. = 1 ft 0 in.

A steel elbow is to be made as shown in the sketch. The main part of the elbow is a quarter of a torus and the small pipe is round.

1. Show the entire curve of intersection in the front elevation.
2. Lay out a pattern that could be used for wrapping around and marking the round pipe so that it could be cut to fit exactly on the elbow. It could then be welded in place.

Group 15. Offsets and combination surfaces

9.15.1.   Scale: ¼ in. = 1 ft 0 in.

Pipes *A* and *B* are both level and both 6 ft in diameter. The axis of pipe *B* is 2 ft 2½ in. higher than the axis of pipe *A*. The two pipes have to be connected by an offset piece as shown.

1. Find the true shape of the right section of the offset piece. Omit the front elevation and use the right side elevation instead.
2. Show the development of the offset.

9.15.2.   Scale: ¼ in. = 1 ft 0 in.

Use the same centerline layout as for Prob. 9.15.1. The pipe *A* is 6 ft in diameter and the pipe *B* is 5 ft in diameter.

Both axes are level. The connection will now be a conical offset.

Show the development of this conical offset.

9.15.3.   Scale: 1 in. = 1 ft 0 in.

Offset connection in a grain elevator between the rectangular bottom of a hopper and the top of a round pipe.

Show the complete development of the offset piece.

The surface is shown divided into four planes and four quarters of cones. Each quarter of the circle is a quarter of the base of a cone having a vertex at one of the four corners of the hopper.

9.15.4.   Scale: ⅜ in. = 1 ft 0 in.

Smoke-pipe transition from a round to a square stack.

Show the development of one quarter of the transition. The rest is similar.

Follow the suggestion offered in Prob. 9.15.3 regarding dividing up the surface into cones and planes.

Group 16. Miscellaneous

9.16.1.   Scale: ⅛ in. = 1 ft 0 in.

The points *A*, *B*, and *C* are three located points on the face of a dam.

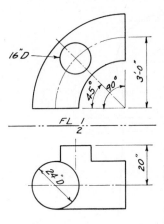

PROB. 9.14.13

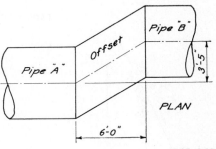

PLAN

PROB. 9.15.1

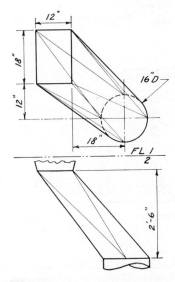

PROB. 9.15.3

A is 4 ft west and 17 ft north of *C*, and at an elevation of 1,826 ft. *B* is 16 ft east and 4 ft north of *C*, and at an elevation of 1,823 ft. *C* is at an elevation of 1,810 ft.

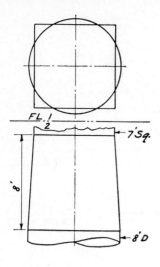

PROB. 9.15.4

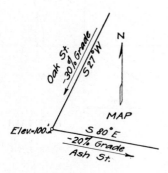

PROB. 9.16.2

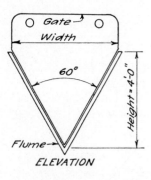

PROB. 9.16.3

A is the center of an opening 12 ft in diameter and running perpendicular to the face of the dam for a depth of 4 ft.

B is the center of an opening 12 ft square and running perpendicular to the face of the dam for a depth of 8 ft. The square opening is also boarded

out from the face of the dam for a distance of 4 ft. Two sides of the square opening are level.

Show the two openings in the plan and front elevation views.

9.16.2.   Scale: 1 in. = 10 ft.

The drawing shows the map of a corner lot. It is desired to level off this lot at an elevation of 110 ft so that each sloping bank will have a 1:1 slope from the level top to the property line.

Show where the limiting edge of the level part of the lot would be and where the two sloping banks would intersect.

9.16.3.   Scale: ¾ in. = 1 ft 0 in.

A triangular-shaped flume and a steel gate. One gate for controlling the water flow in this flume was shipped the correct height but the bottom angle was made 50° instead of 60°.

A second gate was shipped the correct height but the bottom angle was made 70° instead of 60°.

Show how both gates could be used temporarily without alteration until the correct gate arrived.

9.16.4.[L]   Scale: ¼ in. = 1 ft 0 in.

The centerlines of two 8-in. pipes are AB and BC, and the pipes meet at B.

B is 13 ft due east of A, and 6 ft below A.

C is 6 ft north and 6 ft east of A, and 2 ft above A.

From B a 12-in. pipe continues down, so that its centerline has a 15° slope and makes equal angles with AB and BC.

Find:

1. The angle between the two 8-in. pipes.
2. The true bearing of the 12-in. pipe.
3. The angle between the 12-in. pipe and the plane of the 8-in. pipes.

Show a freehand sketch of the special flanged fitting required at B, indicating on the sketch the values in questions 1 and 3.

9.16.5.[L]   Scale: 1 in. = 50 ft.

A penstock conducts water from the forebay to the powerhouse. From A the pipe is level through the forebay wall.

Find:

1. True length of the centerline from A to B and from B to C.
2. True angle (in degrees) of the pipe bends at A and B.
3. True angle between the part BC and the wall of the powerhouse.
4. True shape of the hole in the powerhouse wall.

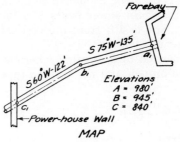

PROB. 9.16.5

9.16.6.   Scale: 1 in. = 40 ft.

A pipeline crossing a canal.

Using the centerlines only, find:

1. The true length of each of the three portions shown.
2. The true size of the four bend angles at A, B, C, and D.

9.16.7.   Scale: ½ in. = 1 ft 0 in.

Trap and sewer problem, as taken from a nationally known trade magazine.

The stock trap and stock branch are as shown.

The vertical centerline of the trap is a fixed distance away from the centerline of the sewer as shown in the plan.

AB is the main sewer, which bears due east, and it has so little slope that it may be considered level. It is desired to determine where to place the branch along the main sewer and the trap along the vertical centerline so that they may be connected by a straight piece of pipe.

Find:

1. The centerline length of the branch pipe from X to Y.
2. The distance Y would be east or west of X.
3. The bearing of the branch line.
4. The elevation of X above the main sewer.

9.16.8.   Scale: ½ in. = 1 ft 0 in.

Use the same trap branch and the same general setting as in Prob. 9.16.7, changing only the relative position of the main sewer and the trap.

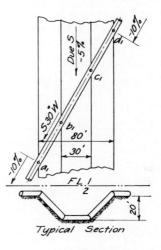

PROB. 9.16.6

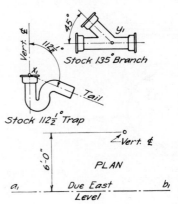

PROB. 9.16.7

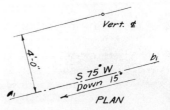

PROB. 9.16.8

For this problem, as shown in Prob. 9.16.8, the main sewer bears S75°W, falls 15°, and is 4 ft from the vertical centerline of the trap.

It is desired to determine where to place the branch and the trap so that

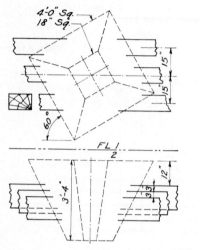

PROB. 9.16.9

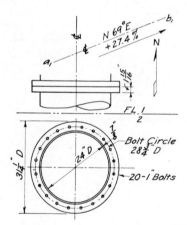

PROB. 9.16.10

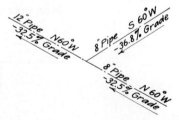

PROB. 9.16.11

they may be connected with a straight pipe.

Find:

1. The centerline length of the branch pipe from X to Y.
2. The distance Y would be south and west of X on the map.
3. The bearing of the branch pipe.
4. The elevation of X above Y to give the height at which the trap will have to be placed.

9.16.9.[T]   Scale: ¾ in. = 1 ft 0 in.

Three 8- by 12-in. floor beams are located as shown, with the 12-in. faces vertical.

The beams are all level but the floor is inclined. The given hopper is shown with a dash line.

Show where to cut the beams so as to allow the hopper to set down through them until the top of the hopper is 12 in. above the top of the highest beam.

Turn the hopper at the angle shown.

Show the hopper with a fine dash line just to indicate where it would go.

Show the beam cuts in both views and crosshatch the visible portion of the cuts.

Solve by one method and check by another method.

9.16.10.   Scale: 1 in. = 1 ft 0 in.

An elbow is required to connect the given level pipe running due north with the pipe AB which bears N69°E and rises on a 27.4 per cent grade.

1. Find the true angle of the bend for the elbow.
2. Locate the bolt holes in the connecting flange of the elbow so they will coincide with the bolt holes as shown on the level pipe flange. Measure the angle the centerline must be shifted.
3. Make a detail shop drawing of the elbow.

9.16.11.   Two 8-in. pipes going into a 12-in. pipe are located as shown.

Design a special flanged Y lateral for this condition.

The bolt holes are to be drilled so as to allow standard valves to be at-

	8-in. pipe	12-in. pipe
Thickness of pipe........................	0.46	0.54
Thickness of flange.....................	1⅛	1¼
Diameter of flange......................	13½	19
Diameter of bolt circle..................	11¾	17
Size of bolt holes.......................	⅞	1
Number of bolt holes....................	8	12
Distance from center to outside face of flange.	15¼	6

tached to the 8-in. or the 12-in. flanges and have each valve stem in a vertical plane containing the centerline of its respective pipe.

The bolt holes on the standard valves straddle the centerlines.

1. Solve all the necessary angles.
2. On a separate sheet make a complete shop drawing of the lateral. Scale: 1½ in. = 1 ft 0 in.

The above information is taken from a table of standard fittings and it is to be used in dimensioning the lateral.

9.16.12.[L]   Scale: ¾ in. = 1 ft 0 in. Belt and pulley problem.

In order for a belt to run on to a pulley it must be moving in the mid-plane of the pulley before it reaches the pulley. With the two pulleys located as shown in the sketch, it is impossible for an endless belt to run in either direction.

You are to install a guide or idler pulley (24 in. in diameter) to guide the belt so that it will operate in either direction over the given pulleys.

Show this new pulley in both the given views. Omit the belt.

Find the arc of contact between the belt and the idler pulley.

All pulleys have a 6-in. face and the crown may be neglected. All shafts are 2 in. in diameter and the belt is 4 in. wide.

Suggestion: Choose any point on the line of intersection of the mid-planes of the two pulleys. From this point draw a line tangent to each pulley. These two lines determine the mid-plane of the required idler pulley.

9.16.13.[T]   Scale: 1 in. = 2 cm.

In nuclear physics, charged parti-

cles moving in a magnetic field travel in a helical path. Two cloud-chamber photographs taken at the same instant at 90° to each other give essentially a plan and elevation view of the path of the particle. Problem 9.16.13, in which dimensions are shown in centimeters, shows three successive positions, A, B, and C, of the same particle taken from photographic exposures. Line XY, the direction of the magnetic lines of flux, lies parallel to the axis of the helix. Find the radius and pitch of the helical path.

9.16.14.   Scale: 1 in. = 30 ft. Culvert problem.

The bearing of a road is due east with a rising grade of 10 per cent toward the east. Station 37 + 60 on the road is just 30 ft above the centerline of a stream flowing S45°W on a 15 per cent grade.

A culvert 8 ft square, outside measurement, is to be so placed that it will carry the stream under the road fill. The centerline of the bottom plane of

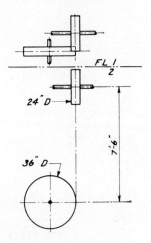

PROB. 9.16.12

the culvert coincides with the centerline of the stream.

The road surface is 25 ft wide. The

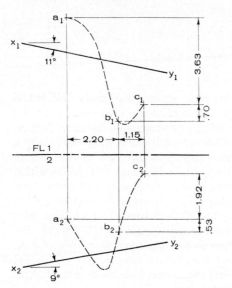

PROB. 9.16.13

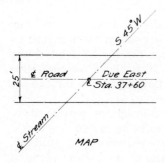

MAP

PROB. 9.16.14

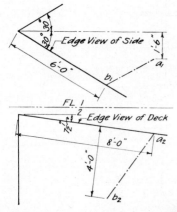

PROB. 9.16.15

side slopes of the road are 1 vertical to 1½ horizontal, but measured in a vertical plane at right angles to the plan view of the centerline of the road. (This gives a slope that is not theoretically correct, but it is the actual way it would be surveyed in road work.)

Find the true length of the centerline of the top plane of the culvert between the two side slopes.

9.16.15.[L]   Scale: ¾ in. = 1 ft 0 in.

The line *AB* is the centerline of a hawse pipe for a boat. The pipe is a cylindrical cast-steel pipe 12 in. inside diameter with walls 1 in. thick. A flange 1 in. thick and 4 in. wide is cast on the lower or outboard end.

1. Make a completely dimensioned shop drawing of the pipe and flange cast in one piece ready to go into place.
2. Make a shop drawing of the deck flange to fit flush against the deck end of the pipe.
3. Make a dimensioned drawing of a template to be used for cutting this pipe out of a stock pipe. The template would be wrapped around the pipe. In this case the flange would have to be welded on.

For the sake of simplifying this problem, assume the small portion of the deck and the sides shown are plane surfaces.

9.16.16.[L]   Scale: 1 in. = 1 ft 0 in.

Connect the two square openings in the walls by the largest possible grain spout made of steel plate. Lay out a development of this spout and dimension it completely. Allow a 2-in. lap all around for riveting (except where two edges come together).

Use the revolution method for finding the true length of all plane intersection lines.

9.16.17.   Scale: ½ in. = 1 ft 0 in.

Lines *AB* and *CD* are the centerlines of two pipes. Show how to connect these two pipes using a 45° branch in the pipe *AB* and a 60° branch in the pipe *CD*. Notice the direction of flow in both pipes. Install the connection so as to maintain the smoothest flow. Find and measure the true length and true slope of this new connection.

9.16.18.[L]   Drift-barrier problem.

The drift barrier used in this problem was actually constructed in a western river to keep driftwood away from the intakes to the water lines. Problem 9.16.18 shows a partial plan view of the 14 piers in which heavy rods were anchored for the purpose of holding the wire barriers. A half elevation of one of the barriers is also shown.

1. Draw a complete plan and front elevation of a typical pier. Scale: $\frac{3}{16}$ in. = 1 ft 0 in. Place these views on the left side of the sheet.

2. In these views show in position the 10 eyebars, each bar lying in a horizontal plane with the vertex of its bend on the line AB (shown in the front elevation). The lowest horizontal plane is 1 ft above low-water elevation and the planes are 1 ft apart. Locate the exact points at which each end of each bar comes through the sloping concrete wall of the pier.

3. Using a scale of $\frac{1}{2}$ in. = 1 ft 0 in., detail one typical eyebar and make a table of lengths for all the bars. Allow each end of each bar to project 6 in. out from the concrete.

9.16.19.   Scale: 3 in. = 1 ft 0 in.

Draw a left-hand screw conveyor with a convolute blade on a 5-in. shaft and inside a 12-in. cylinder. The lead is 8 in. The shaft is level.

Lay out the development for one turn of the blade. Calculate the number of turns of the blade that could be made in one piece of metal.

9.16.20.[L]   Scale: $\frac{1}{2}$ in. = 1 ft 0 in.

Dead-end tower for an electric power line.

This is an actual problem that was given to a draftsman who was working for a power company.

The drawing shows part of a three-pole tower for a 167,000-volt line. The dash lines from a_1 and c_1 to the frame represent strings of insulators between the conductor and the frame.

Between A and C the conductor is allowed to sag in a loop, as shown in the end elevation; this is called a jumper.

It is necessary to have a definite clearance between the nearest points on the live wire and the guy wire. This presents the three conditions which require checking.

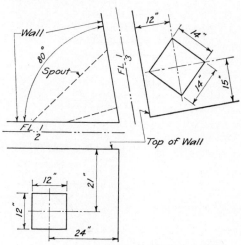

PROB. 9.16.16

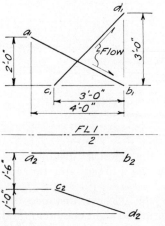

PROB. 9.16.17

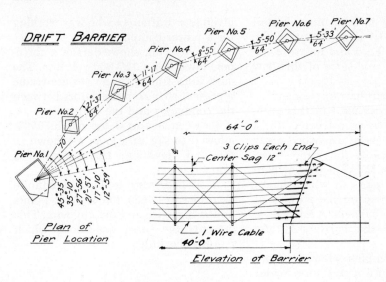

DRIFT BARRIER

Plan of
Pier Location

3 Clips Each End
Center Sag 12"

64'-0"

1" Wire Cable
40'-0"

Elevation of Barrier

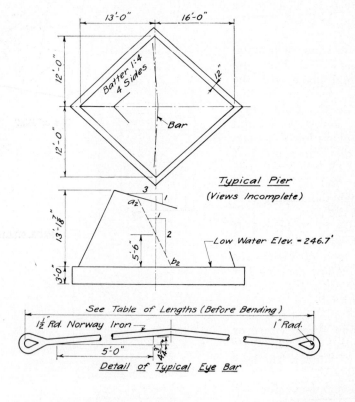

13'-0"   16'-0"

12'-0"

Batter 1:4
4 Sides

12"

Bar

12'-0"

Typical Pier
(Views Incomplete)

13'-7⅛"

3'-0"

5'-6"

Low Water Elev. = 246.7'

See Table of Lengths (Before Bending)

1½" Rd. Norway Iron   1" Rad.

5'-0"

Detail of Typical Eye Bar

PROB. 9.16.18

1. Find the closest distance between the jumper and the guy wire.

2. Find the closest distance between the straight conductor and the guy wire.

3. Find the closest distance between the guy wire and the jumper when the wind blows it 60° away from its vertical position.

Scale all distances accurately. As-

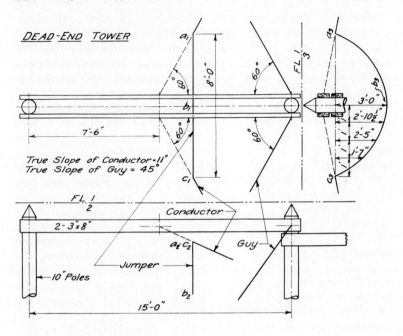

DEAD-END TOWER

True Slope of Conductor-11°
True Slope of Guy = 45°

PROB. 9.16.20

sume that the guy wire and the conductor (except the jumper) are both
straight lines for the short length
which is used.

9.16.21.[L]   Scale: half size.

A small hand-operated concrete
mixer for laboratory use has a horizontal cylinder 12 in. in diameter. The
blades are attached to arms on the
shaft at the angle shown. The blades
are plane surfaces and they have a
true slope of 12 (vertical) to 5 (horizontal).

Make a layout of one metal blade
designed so as to give ⅛ in. clearance
from the cylinder for the entire length
of the blade.

9.16.22.[T]   Scale: ³⁄₁₆ in. = 1 ft 0 in.

The setting for this problem is taken
direct from a problem that arose in the
yards of a transcontinental railway.

The centerline of a track bears due
north and is level. The points A and B,
located as shown, are the centers of
two catch basins.

They are both 12 in. below the centerline of the track.

From A a drainage pipe bears due
east under the track and has a falling

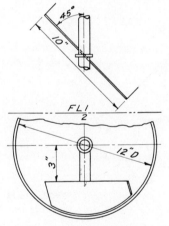

PROB. 9.16.21

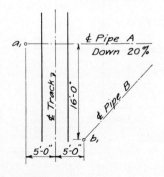

PROB. 9.16.22

grade of 20 per cent. A second drain pipe from B is to connect into the pipe from A.

The engineer wishes you to furnish him the following information:

1. In order to use a stock 45° connection, find the length of the A pipe, the length of the B pipe, and the bearing of the B pipe.

2. In order for the two drain pipes to have the same grade, find the length of the A pipe, the length of the B pipe, the bearing of the B pipe, and the angle for the special connection required.

9.16.23.   Scale: ½ in. = 1 ft 0 in.

This is an actual problem which recently had to be solved by the Bureau of Fisheries. CD is the centerline of the discharge pipe coming from a pump house. AB is the centerline of the main distribution pipe to the ponds. Both pipes are 12-in. pipes and

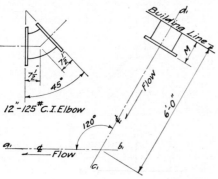

PROB. 9.16.23

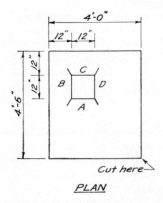

PLAN

PROB. 9.16.25

both centerlines are level. However, the pipe AB is at an elevation 4 ft 10 in. higher than the pipe CD. The problem is to connect these two pipes using two standard 45° elbows as shown in Prob. 9.16.23.

1. Find the distance M for locating the face of the discharge-pipe flange to which the elbow is to be connected.

2. Find the distance between the flanges of the sloping pipe connecting the two elbows.

3. Using the information just determined, on a separate sheet make a complete plan view showing the elbows and the connecting pipe in place. The outside diameter of the pipe is 12¾ in., the flange diameter is 19 in., and the flange thickness is 1¼ in.

9.16.24.[L]   Scale: 1 in. = 20 ft. Two pipe lines, AB and CD.

B is 57 ft south and 42 ft east of A and 22 ft above A.

C is 59 ft south and 5 ft west of A and 2 ft below A.

D is 21 ft south and 40 ft east of A and 15 ft below A.

Locate the shortest possible pipe having a true slope of 45° and connecting the pipes AB and CD. Find the bearing and true length of this pipe.

9.16.25.[L]   Scale: ½ in. = 1 ft 0 in.

A hopper to be made out of sheet metal is shown in the plan view only. The hopper is 6 ft high with a level top and bottom, and the sides are vertical planes until they meet the sloping bottom planes. The bottom hole is 12 in. square. Bottom planes A and C have a slope of 45° and bottom planes B and D have a slope of 60°.

1. Draw complete plan, front and right side-elevation views, including all hidden lines.

2. Make a one-piece layout (development) of this hopper, cutting it at the vertical corner indicated. All other vertical corners are not to be cut, but all the sloping bottom corners will have

to be cut. Place the outside of the metal up. The method of revolution is recommended for finding true sizes of planes.

9.16.26.[L]   Scale: ¼ in. = 1 ft 0 in.

This problem arose in building the Mud Mountain Dam in the state of Washington. A cylindrical intake shaft 12 ft in diameter stands at a slope of 4 to 1. It is to have a conical roof extending part way around the top until it becomes tangent, on both sides, to a gable roof (V-roof) with a level ridge.

1. Show where the gable roof will meet the conical roof.
2. Find the number of degrees the conical roof will extend around the cylinder.
3. Find the slope of the gable roof.
4. Check the slope by calculation.

9.16.27.   See Warner and Douglass Problem Book.

9.16.28.   Scale: ⅜ in. = 1 ft 0 in.

Assume light rays in the direction shown in Prob. 9.16.28. Determine the shadow cast on the steps in both views and also the shadow cast on the level ground.

9.16.29.   Scale: 1 in. = 1 ft 0 in.

Using light rays as shown in Prob. 9.16.29, work out the shadows cast by the round column and its cap. Find the shadows cast on the floor, the wall, the step, and the column itself. Dash any shadow lines that are invisible.

9.16.30.[L]   Scale: 1 in. = 300 ft.

Point B is 900 ft east and 100 ft north of A and 250 ft higher than A. Point C is 400 ft east and 600 ft north of A and 450 ft higher than A.

In order to send a microwave beam from A to B, it is necessary to install a reflector at C. Point C is to be the center of the reflector, which is rectangular in shape. Its size is 6 by 8 ft, and the 6-ft edges are to be level.

Show the reflector in all views, using a blown-up scale of 1 in. = 4 ft. Measure and record the data required for installation, such as the true slope

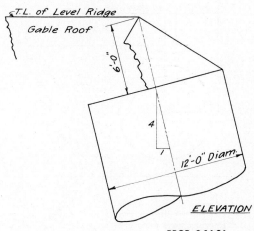

ELEVATION

PROB. 9.16.26

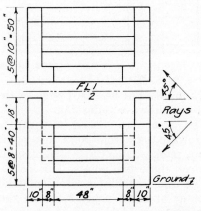

PROB. 9.16.28

of the reflector and the bearing of its level edges. Also measure the angle the beam makes with the reflector.

9.16.31.[L]   Use the same scales and the same data as for Prob. 9.16.30 except that the reflector to be installed must be round instead of rectangular. Make the reflector 10 ft in diameter and show it only in the plan and front elevation views. Use the axes and trammel method. Measure and record the true slope of the reflector and the bearing of a level diameter. Also measure the angle the beam makes with the reflector.

9.16.32.[L]   Layout scale: 1 in. = 1,000 ft. Detail scale: 1 in. = 10 ft.

The following data are given exactly as used recently by a power com-

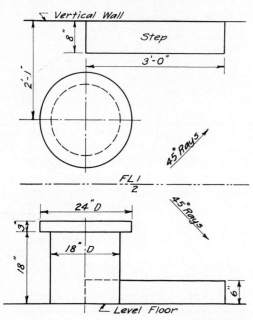

PROB. 9.16.29

pany in the Pacific Northwest to install a microwave radio link.

The company wished to have a microwave path from point *A* to point *C*, but there was no direct line of sight between these two points. This necessitated the installation of a metallic reflector at some point *B* where the microwave could be reflected from *A* to *C*. This reflector must have 169 sq ft of projected area measured perpendicular to the lines of sight to *A* and to *C*. The long sides of the reflector must be parallel to the plane of the two lines of sight.

Point *B* is located N34°34′W from *A* and at an elevation of 2,020 ft. The true length of the line of sight from *A* to *B* is 2,660 ft, and it has a true slope rising toward *B* of 32°43′.

Point *C* is located N59°58′E of *B* at an elevation of 2,515 ft. The true length of the line of sight from *B* to *C* is 4,875 ft.

1. Find the dimensions of the reflector, using the detail scale.
2. Show the reflector in the plan view, with a level line through its mid-point.

3. Measure the bearing of this level line.
4. Find the true slope of the plane of the reflector.

Group 17. Mining problems

9.17.1.   Scale: 1 in. = 50 ft.

A, *B*, and *C* are known points of outcrop on a vein of ore.

B is 125 ft due south of *A* and 100 ft below *A*.

C is 100 ft south and 100 ft east of *A* and 25 ft below *A*.

Find:

1. Strike of the vein.
2. True dip of the vein.
3. True slope in going from *A* to *B* direct.

9.17.2.   A vein of ore is cut off by a vertical cliff which runs due east and west. The streak of the exposed ore as it is seen on this cliff appears to dip down 52° in an easterly direction. The strike of the vein is known to be N45°E. Find the true dip of the vein.

Note: An easterly dip does not mean that a vein dips exactly due east. It does mean that the vein is low in the eastern portion of the map and high in the western portion. Or, any point on the vein is lower than any other point on the vein from which it is due east.

9.17.3.   An inclined mine shaft *AB*, and a vertical shaft. Scale: 1 in. = 30 ft.

A is 60 ft north and 56 ft west of *B* and 40 ft above *B*. The vertical shaft is 35 ft due north of *B*. The bottom, *C*, of the vertical shaft is 45 ft higher than *B*.

Find:

1. The bearing, true length, and per cent grade of the shortest shaft to connect *C* with *AB*.
2. The bearing, true length, and per cent grade of the steepest shaft to connect *C* with *AB*.

9.17.4.   Scale: 1 in. = 100 ft.

A is a point of outcrop of ore. A

borehole is started at M, which is 150 ft west and 125 ft south of A and 25 ft below A. The hole bears S60°E and falls 45°. Ore is struck after boring 130 ft. A second hole is started at N, which is 150 ft east and 75 ft south of A and 25 ft below A. This hole bears S45°W and falls 60°. Ore is struck after going 190 ft. Neglect the thickness of the vein.

Find:

1. Strike of ore vein.
2. Dip of ore vein.

9.17.5.   Scale: 1 in. = 10 ft.

A vein of ore whose strike is S60°E is known to be 9 ft thick. A level tunnel bearing due south from A pierces the upper surface of this vein after going 15 ft and the lower surface after going an additional 25 ft. If the tunnel from A had been driven with the same bearing and a -30 per cent grade, what would have been the distance to the vein?

9.17.6.   Scale: 1 in. = 10 ft.

Three short holes are drilled through a vein of ore from a point of outcrop A on the upper surface of the vein. One hole bears due north on a rising 20 per cent grade and shows 20 ft of thickness to the vein; a level hole bears due west and shows 10 ft of thickness; the third hole is vertical and shows 15 ft of thickness.

Find the strike, dip, and real thickness of the vein.

9.17.7.   Scale: 1 in. = 40 ft.

A vein of ore is located by three points of outcrop, A, B, and C. B is 100 ft due north of A and 80 ft higher than A. C is 90 ft due east of B and 60 ft lower than B. A vertical hole drilled at A shows the vein to be 25 ft thick.

Find the strike, dip, and real thickness of the vein.

9.17.8.   Scale: 1 in. = 20 ft.

A and B are two points on the upper surface of a vein of ore and X and Y are two points on the lower surface.

B is 50 ft east and 35 ft south of

A and is also 25 ft higher than A.

X is 60 ft due east of A and 30 ft higher than A.

Y is 30 ft due east of A and 5 ft lower than A.

Find the strike, dip, and thickness of the vein.

9.17.9.   Scale: 1 in. = 40 ft.

A vein of ore is known to have an easterly[1] dip and to be 20 ft thick. A level borehole bears due west from A and intersects the upper and lower faces of the vein at 35 ft and 70 ft, respectively, from A. X is a point of outcrop on the upper face of the vein and is 50 ft due north of A and 25 ft higher than A.

Find the strike and dip of the vein.

9.17.10.   Scale: 1 in. = 20 ft.

A vertical borehole from A intersects the upper surface of a vein at a depth of 8 ft and the lower surface at a depth of 30 ft from A. A second borehole is sunk from a point B which is 28 ft due north of A and 13 ft lower in elevation than A. This hole bears due east, falls 60°, and intersects the upper surface of the vein at 30 ft and the lower surface at 50 ft, distances being measured from B along the hole.

Find the strike, dip, and thickness of the vein.

9.17.11.   Scale: 1 in. = 20 ft.

A vein of ore is known to have a westerly[1] dip and a thickness of 14 ft. Two vertical boreholes are sunk at the points A and M. A is 40 ft west and 23 ft north of M and on the same level as M. The hole at M reached the vein at a depth of 10 ft and the one at A at a depth of 30 ft. Both holes showed a thickness of 20 ft through the vein.

Find the strike and dip of the vein.

9.17.12.   See Warner and Douglass Problem Book.

9.17.13.   Scale: 1 in. = 60 ft.

A fault plane has a strike due east and a southerly dip. The rake of the slickensides is known to be 45°, and

[1] See note under Prob. 9.17.2.

the surveyed bearing of the slickensides is S60°E.

A vein with a strike due north is
offset along the fault strike, its strike
(at the same level) being 160 ft far-
ther east on the south wall. The net
slip is 140 ft.

Find and measure the dip of the
fault, the dip of the vein, and the
plunge of the slickensides.

9.17.14.[L]   Scale: 1 in. = 60 ft.

A fault has a strike due east and
dips 60° south. A vein striking N30°W
and dipping 45° toward the south is
offset 200 ft along the strike of the
fault. The rake of the slickensides on
the fault is known to be 50° in a
southeasterly direction.

Find the following and record
where measured:

1. Net slip.
2. Horizontal component of net
   slip.
3. Vertical component of net slip.
4. Bearing of net slip.
5. Plunge.

9.17.15.[L]   Scale: 1 in. = 200 ft.

A fault plane has a strike due east
and dips 60° north. Three veins, dis-
rupted by the fault, have strikes and

dips as shown in the illustration.

Find the following and record results where measured:

1. Net slip.
2. Plunge.
3. Bearing of net slip.
4. Horizontal component of net
   slip.
5. Vertical component of net slip.
6. Rake.
7. Location of vein C on the south
   side of the fault, as determined
   by its strike line.

Note: The strike lines of the veins
and the fault plane are all taken at
the same elevation.

Group 18. Aeronautical problems

9.18.1.[L]   Scale: full size.

The mid-plane circles of two pulleys
in an airplane are located as shown.
The cable must operate over these two
pulleys moving in either direction.
When moving one direction the cable
enters pulley B and leaves pulley A,
as shown by the arrows.

It is desired to install a third idler
or guide pulley to the left of the bulkhead so the cable can operate in either
direction. All three pulleys are 1½ in.
in diameter and it is only necessary
to show the mid-plane of each pulley.

See Prob. 9.16.12 for an explanation of the principle of belt or cable
operation and for a suggestion as to
the solution. Choose the point on the
intersection of the two planes about
1¼ in. to the left of the bulkhead so
the pulley will clear the bulkhead.

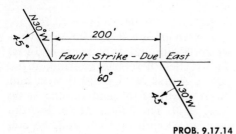

PROB. 9.17.14

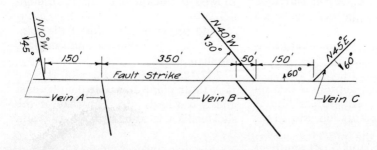

PROB. 9.17.15

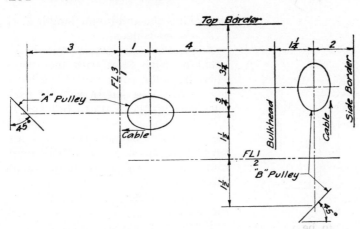

PROB. 9.18.1

1. Show the idler pulley in the plan view.
2. Find, in degrees, the arcs of contact between the cable and each pulley and show these arcs in the plan view.

9.18.2.[L] Scale: $\frac{1}{10}$ size. *AB* and *CD* are two control cables in an airplane.

1. Find the clearance between these two cables at the closest point.
2. If the clearance must be a minimum of 4 in., how far would you have to lower the point *B* vertically to provide this minimum clearance between cables? The point *B* is the only point which may be moved.

9.18.3.[L] Scale: 3 in. = 1 ft 0 in.

The following problem is almost an exact duplicate of an actual problem that had to be solved recently in fitting a tube against the aileron of an airplane. The data have been changed slightly just for convenience in making the solution.

A level metal tube must be cut so it fits exactly flush against the sloping plane *ABCD*, shown in Prob. 9.18.3. Find the angle at which the pipe must be cut off. Make a development of the pipe which may be used as a template to wrap around the pipe for cutting the end next to the plane.

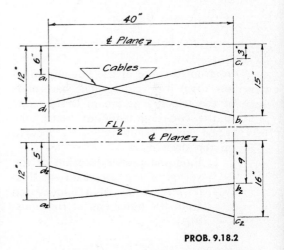

PROB. 9.18.2

A lug, shown on the outer end of the pipe, is in the same vertical plane as the centerline of the pipe. Locate the centerline of this lug on the development so the lug will remain in the correct position after the pipe is in place.

9.18.4.[L] Scale: $\frac{1}{4}$ size. Locate the thrust centerline close to the bottom of the sheet.

Tripod support bracket for a starter extension shaft.

The legs of the tripod are tubes $\frac{1}{2}$ in. outside diameter and the engine-mount tubes are 1 in. outside diameter. At the vertex of the tripod the tubes are to be welded to the bearing housing and they must be at least $\frac{1}{4}$

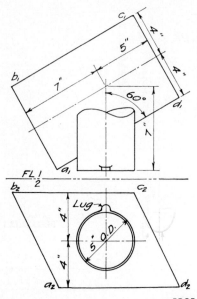

PROB. 9.18.3

in. from the end of the housing to allow room for welding. The base ends
of the tripod legs are to be fastened
to the two engine-mount tubes, one
tripod point being definitely located
on the drawing.

1. Find the proper location of the
   vertex of the tripod on the centerline of the shaft, measured
   from the outer end of the housing.

2. Find the minimum length A of
   the housing.
   Suggestion: The vertex must be located by trial. Start with the vertex
   ½ in. from the outer end of the housing. In the true length view of each
   leg, blow up the scale to full size in
   order to check the minimum edge
   clearance of each tube. Then, if necessary, shift the vertex until the minimum clearance is maintained for all
   three tubes.

   9.18.5.[L]   Scale: ⅟₂₀ size. Start on
   the left.

   Retractable landing gear.

1. Locate the centerline of a shaft
   (or axis) about which the wheel
   may be revolved from the down
   position to the retracted position inside the wing as fixed on
   the drawing. The point B is any
   point on the wheel axle, preferably not less than 20 in. from A.

2. Find the number of degrees
   through which the wheel will
   rotate about this axis to reach
   the retracted position.

   9.18.6.[L]   Scale: ⅟₂₀ size. Start on
   the left.

   Retractable landing gear with
   skewed axis.

   The points A, B, C, and D define a

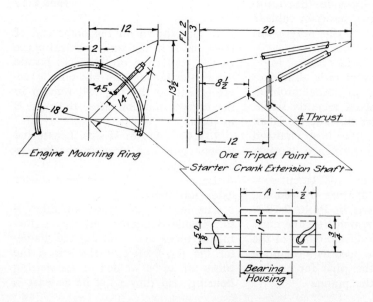

PROB. 9.18.4

portion of the lower surface of a wing having a standard NACA 2418 section. The sweepback angle is 7°, the angle of incidence is 3°, and the dihedral angle is 5°. Using the data as given above, the exact locations of A, B, C, and D have been calculated mathematically and are as given on the drawing.

A skewed axis is to be located about which the wheel may be retracted to the position shown inside the wing, with the lower face of the wheel parallel to and 1 in. from the lower wing surface. The point R is a point on the oleo strut centerline through which the skewed axis must pass and has been assumed far enough from the leading edge of the wing to clear the front wing spar. The entire oleo centerline, when retracted, will lie 5 in. from the surface ABCD, assuming this small portion to be a plane. Therefore the point R is also 5 in. from the wing surface and its elevation must be

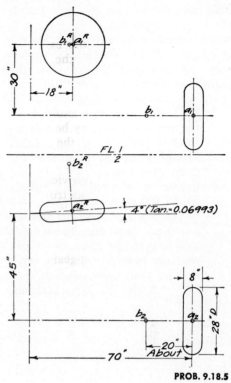

PROB. 9.18.5

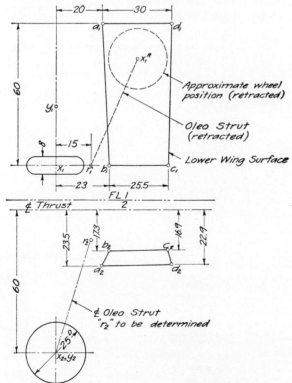

PROB. 9.18.6

determined to obtain the true length of the oleo strut XR.

The point Y is any point on the centerline of the wheel axis, preferably not less than 25 in. from the point X at the wheel center. This point is the key to the solution.

1. Find the retracted position of the wheel. Locate the exact retracted position of the points X and Y in the plan and the front elevation.

2. Locate a skewed axis about which the wheel may be rotated to place X and Y at the desired retracted position. Show this axis in both the given views and measure in degrees its true slope and its plan view position.

9.18.7.[L]   Scale: ⅒ size. Start on the lower left.

Double tripod landing-gear structure.

The three wheel loads on this structure are assumed to act through the vertex V of the lower tripod ABC. The load on the member A will be the imposed load on the upper tripod EFG.

The given wheel loads are:

$$V = 8,800 \text{ lb}$$
$$S = 2,630 \text{ lb}$$
$$D = 3,310 \text{ lb}$$

Using a vector scale of 1 in. = 2,000 lb, find graphically the stress in each member of the lower tripod only.

Suggestion: Use a view showing members B and C as an edge.

9.18.8.[L]   Scale: ½₀ size. Start on the lower left.

Find graphically the stress on each member of the upper tripod EFG shown in Prob. 9.18.7.

Assume the imposed load through the member A to be 7,135 lb C.

Vector scale: 1 in. = 2,000 lb.

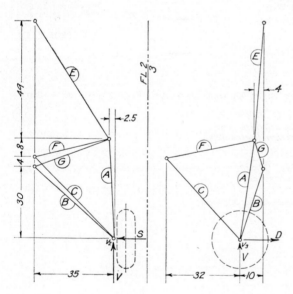

APPENDIX

A.1 **Procedure for drawing parallel and perpendicular lines not square with the margins of the paper**

Parallel and perpendicular lines that are square with the margins of the paper are drawn with the T square or T square and triangle in combination. When the lines are not square with the edges of the paper, as is frequently the case in descriptive ge-

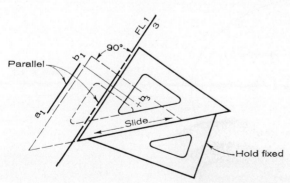

FIG. A.1. Drawing parallel and perpendicular lines.

ometry problems, other methods must be used. It is wise to learn how to do this operation efficiently and accurately.

The 30–60° triangle and 45° triangle are used in combination, sliding them on their *hypotenuses*, as illustrated in Fig. A.1, with the 45° triangle *uppermost*. This arrangement gives the maximum sliding range and provides equal lengths on either side of the 45° triangle for drawing parallels or perpendiculars.

In Fig. A.1, folding line 1-3 may be drawn parallel to a_1b_1 and the projection line from b_1 to b_3 drawn perpendicular to the folding line by sliding the upper 45° triangle while holding the lower triangle fixed with the fingers of the left hand.

A.2   Trammel method for drawing an ellipse

An ellipse may be quickly and accurately drawn by the use of a trammel when the lengths of the major and minor axes are known. The best trammel to use is a thin card or paper with a very straight edge. Along this straight edge mark some point A, as in Fig. A.2, which is the point which will draw the ellipse.

Mark the point B at a distance of half the minor axis from A. Also mark the point C at a distance of half the major axis from A. Lay the trammel across the two axes as shown so the points B and C touch the axes and make a pencil mark on the drawing opposite the point A. This will be one point on the ellipse. In the same manner lay the trammel in several other positions until sufficient points are marked to establish the entire curve. Then draw the ellipse with a French curve.

A.3   Sheet-metal workers' method for finding true lengths of elements

Sheet-metal workers use a special short-cut method for finding the true lengths of several elements in preparation for develop-

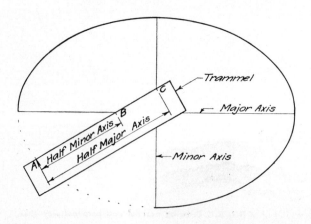

FIG. A.2. Trammel method.

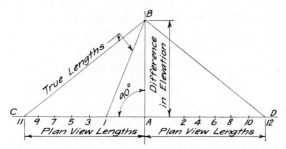

FIG. A.3. Special method for true lengths. Same difference in elevation.

ing a curved surface. This method is well worth knowing because it is convenient and accurate and because it saves much
confusion on the drawing.

Case 1.   When all the elements have the same difference in
elevation between their two ends. See Fig. A.3.

Lay out the lines *AB* and *CD* at right angles to each other.
Make *AB* equal in length to the difference in elevation between
the two ends of any element. From *A* measure along the line *CD*
the plan-view length of each element and number each point thus
obtained with the element number. These points may be measured to the left or to the right of *AB*. For convenience in Fig. A.3,
the even-numbered elements are shown on the right and the odd-
numbered on the left. From *B* to these ends just numbered along
CD are the true lengths of the elements, which do not actually
need to be drawn. The true lengths of all the elements needed
are on this diagram and may be transferred to the development
with dividers.

Case 2.   When all the elements do not have the same difference in elevation between their two ends. See Fig. A.4.

The general procedure is the same as for Case 1 except that
the differences in elevation, since they are all different, are laid
out from *A* toward *B* and must also be numbered with the num-

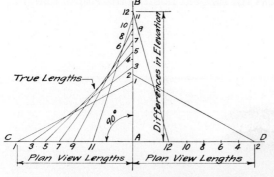

FIG. A.4. Special method for true lengths. Not the same difference in elevation.

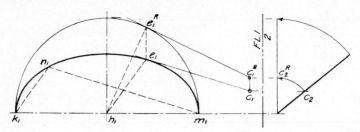

FIG. A.5. Line tangent to an ellipse.

ber of the element. The distances from the numbered points along AB to the corresponding numbered points along CD are the true lengths of the elements.

A.4   **To find exact point where a straight line is tangent to an ellipse**

In Fig. A.5 the problem is to locate the exact point where a straight line from point C is tangent to the ellipse. The point C must be in the same plane as the ellipse.

Case 1.   Without using the circle.

1. Draw the line from c_1 tangent to the ellipse.

2. Draw $m_1 n_1$ parallel to this tangent.

3. Draw $k_1 n_1$.

4. Draw $h_1 e_1$ parallel to $k_1 n_1$. This determines the exact point of tangency at e_1.

Case 2.   By using the circle.

Any ellipse may be considered to be the projection of some circle.

1. Construct this circle and revolve it down to a position in view 2 where it projects as the given ellipse.

2. Project point C to the plane of the circle in view 2 and then revolve the whole plane to a level position. Point c_1 moves to c_1^R.

3. Now draw the tangent from c_1^R to the circle and locate the exact tangent point at e_1^R by the perpendicular $h_1 e_1^R$.

4. Revolve the plane back to its slanting position, and e_1^R moves to e_1. This checks the tangent point to the ellipse found by the method of Case 1.

INDEX